R

Research Report No. 80

Gunilla Andræ & Björn Beckman

Industry Goes Farming

The Nigerian Raw Material Crisis and the Case of Textiles and Cotton

The Scandinavian Institute of African Studies
Uppsala 1987

ISSN 0080-6714
ISBN 91-7106-273-4

Printed in Sweden by
Motala Grafiska
Motala 1987

Contents

Preface

This report is a revised and expanded version of a paper presented to the "Seminar on Nigerian Economy and Society since the Berlin Conference", organised by the Faculty of Arts and Social Sciences, Ahmadu Bello University, and the Nigerian Institute of Social and Economic Research (Ibadan) in Zaria, 11-15 November 1985. The research was undertaken when Gunilla Andrae was a research associate of the Centre for Social and Economic Research at A.B.U. and Björn Beckman was teaching in the Department of Political Science of that university. The study has received financial support from the Swedish Agency for Research Cooperation with Developing Countries (SAREC). It is part of a wider project "Labour and industrial crisis in the third world: the case of Nigerian textiles and cotton". (For an outline, see Andrae and Beckman 1984.)

We wish to acknowledge all help and cooperation received from the staff of the Nigerian Textile Manufacturers' Association, the Nigerian Textile, Garment and Tailoring Employers Association, the National Union of Textile, Garment and Tailoring Workers of Nigeria, the Textile Institute of Nigeria, the Cotton and Agricultural Processors Ltd., the Nigerian Cotton Board, the Institute for Agricultural Research at A.B.U., the Upper Benue River Basin Development Authority, and from the management of a range of individual textile firms, including Afcott, the first industrial cotton growing company. Interviews with individual officials are referenced in the text and listed in the bibliography.

We also wish to acknowledge gratefully the continued support from Professor A.D. Yahaya, at the time of our research the Dean of the Faculty of Arts and Social Sciences at A.B.U. A number of colleagues have contributed with critical and helpful comments on the original presentation and as part of an ongoing debate over peasants and plantations, including Akin Fadahunsi, Rigobert Ladipo, and Gavin Williams.

We welcome comments and reactions. Please write us under the following addresses:

Gunilla Andræ, Department of Human Geography, or Björn Beckman, Department of Political Science, both of University of Stockholm, S-106 91 Stockholm, Sweden.

Stockholm, September 1987.

Introduction

This study has two different purposes. Firstly, we discuss the raw material crisis of the Nigerian textile industry in the mid-1980s, how it came about, and the strategies pursued by industry in dealing with it. The crisis erupted as a sharp fall in domestic cotton production coincided with an equally dramatic decline in national earnings from oil exports, thus making the importation of cotton and other raw materials difficult. In this respect the report is part of a wider ongoing study of the problem confronted by the Nigerian textile industry, especially as seen from the perspective of trade unions and workers (Andrae and Beckman 1984). The raw material crisis had serious consequences for capacity utilisation, employment and workers' income.

Secondly, the study addresses general issues relating to the relationship between industry and agriculture, and in particular the manner in which industry influences social organisation at the agricultural end. In this respect the focus is on industrial strategies for controlling, transforming or replacing peasant producers. It links up with our earlier work on agro-based industries (Andrae and Beckman 1982 and 1985; Andrae 1983 and Beckman 1985b). We argue that there is evidence of a critical new conjuncture where import-dependent industries are "domesticated", are forced to integrate backward into agriculture. We believe it to be a process of major significance for accumulation, class formation, and national development generally. The direct intervention of industrialists into agricultural production ("Industry goes Farming") is only the most conspicuous form taken by this new development. It is one among several simultaneous and competing strategies, including involvement in marketing and "contract farming". These are also discussed.

AN OUTLINE OF THE REPORT

The study is organised as follows. The first chapter suggests some provisional evidence of this new phase in the relationship between industry and agriculture, mostly press reports on direct farm investments by manufacturing companies, not just textile firms, but breweries, flour mills, and the producers of soap, soft drinks, pharmaceuticals and other goods with a content derived from agriculture. This development is set against the background of a long-standing controversy over plantations versus peasant farming, where the social forces favouring plantations were long constrained but lately have been given a major boost. It is also placed in the context of theoretical arguments about industrialisation in the third world, the problem of "disarticulation", the non-integration of economic sectors and the limitations of import-substitution.

The second chapter introduces the Nigerian textile industry and the trends toward raw material shortage. It is based on an examination of industry data from company reports, official statistics, interviews with managements and branch organisations. Changes in the balance between spinning and weaving in the development of the industry and in the balance between man-made and natural fibres all affect the manner in which the raw material crisis makes itself felt. Shifts in consumption patterns and heavy competition from smuggled goods must also be taken into account. The problem, we argue, was not originally shortage of domestic cotton. On the contrary, there was more domestic cotton in the mid-70s than the industry was prepared to handle, due to limitations on domestic spinning capacity. Rather than expand spinning the industry preferred to import spun yarns. Part of this was synthetic yarns to satisfy a growing demand in this direction. But to an equal degree expansion based itself on imported cotton yarns, thereby sidetracking the need to develop and sustain domestic cotton supply. This was possible as the years around 1980 were characterised by virtually unrestricted access to imported raw materials, fibres as well as yarn, at prices much below domestic equivalents.

The development of domestic raw material supply is discussed in chapter III. We outline the rise and fall of Nigerian cotton production and discuss causes of decline, including the shift into food production. Cotton faced special problems in these conjunctures being a labour intensive crop and thus faced with the turbulence of labour markets caused by the ups and downs of the petroleum economy. Since colonial days, cotton had been subjected to state marketing, primarily for tax purposes but also as a means for crop development, seed distribution and the provision of extension services. By the mid-70s, the marketing board system had lost its fiscal clout and was incapable of handling the problems of declining production. Government controlled producer prices failed to match fast rising production costs and competition from a booming food market. The buying organisation disintegrated. The pressures for reform which might have been expected to have come from the textile industry itself were preempted as imported raw materials were allowed to flow freely. Imports therefore contributed to the sharp decline in domestic production.

In 1986, a year after this report was first drafted, the marketing boards were abolished. We have added a postscript to account for this momentous event and to discuss its implications for cotton. Our information on the first cotton season without marketing board (1986/87) is, however, incomplete. The season was characterised by much confusion about actual buying arrangements and by *ad hoc* interventions by state governments. It is therefore difficult to tell what will happen once farmers, traders,.and manufacturers find their feet in the new situation. The picture is further complicated by Nigeria's simultaneous change of foreign exchange regime including a sharp devaluation, also discussed in the postscript. The full impact of that upheaval also remains to be seen.

Chapter IV discusses the strategies pursued by manufacturers in the face of the raw material squeeze. While the struggle continued to rage over preferntial treatment in access to imports and foreign exchange, domestic sourcing was forced on manufacturers, whether by government policies or by the logic of the new foreign exchange situation. Since the imports to be replaced consisted very largely of spun yarns, substitution by local fibre requires the reactivation and expantion of the capacity to spin. New investments in the spinning industry have also been announced. Concern has also been stepped up for developing the capacity to process synthetic materials, when ready to be supplied by the petrochemical industry in the 90s.

The movement into plantation agriculture has already been pioneered by a new set of commercial farmers. Political pressures from this class had removed some important barriers which existed in the past, including restrictions on direct foreign investment and access to land. The Nigerian Textile Manufacturers' Association entered into negotiations with the government and agreed on an import substitution plan for cotton. It committed itself to a programme for cotton rehabilitation for which it taxed its members and provided grants to governments in cotton producing states. The Association also funded research, especially on mechanised harvesting, a technology only relevant for large-scale plantations. It offered to take over key functions of the Cotton Board, the state marketing organisation. Individual companies opted for their own solutions. While some went into direct production others aimed at a system of contract farming, modelled on the successful Nigerian Tobacco Company.

The first substantive investments by individual textile firms in cotton plantations were made in 1986, after we had prepared the draft report. We have added a brief report on a visit we made that year to Afcott's 2 000 ha plantation in Gongola. We suggest that while the plantation is unlikely to produce cotton more efficiently than peasant farmers, it may still be a profitable investment in that it ensures company access to cotton in a situation of extreme shortage. The plantation also provides a platform for alternative strategies for controlling supplies, including forms of cooperation and contractual relations with other producers.

In conclusion we attempt to summarise the logic of the processes observed. Import dependent industrialisation, we argue, leads into a dead end, a "raw material trap" that forces industry to integrate backward. There is particular scope for this in the case of industries with a potential domestic agrarian base. Much third world industry has in fact been established with the pretence of developing such potential. It has not been realised as long as the easier option of importing was still open, especially as foreign owners have a direct stake in the importing business.

Plantations may be profitable to individual owners, especially as strategies of controlling supplies. We argue, however, that they are wasteful in terms of national resources. They cannot be generalised as a strategy, especially as they continue to depend on scarce imported inputs. They are diversionary in that they misdirect resources away from the peasant sector which has the largest potential for productivity increases and income generation.

We draw attention to the international and domestic class interests which promote plantations. The "captured" industrialists provide ideal partners for transnational agri-businesses to whom peasants are poor customers. The new industrial farmers forge alliances with an emerging agrarian bourgeoisie anxious to remove the remaining obstacles to agrarian capitalism, and secure preferential treatment by the state. Plantations constitute one end of a continuum of forms whereby industry seeks to establish control over agriculture. With contract farming firms seek to control individual small producers. Outgrowers' schemes combine plantation and contract components.

We conclude by discussing the prospects provided by the development of a petro-chemical industry as an alternative source of raw materials for the textile industry. The shift into man-made fibres in the past has been as much a cause as a result of the decline of cotton. We argue that chips for synthetic fibres consume massive investment funds and non-renewable resources while providing little employment. Peasant-based cotton production, on the other hand, can provide work and income for vast numbers of rural producers while drawing primarily on renewable resources. Peasant income is vital if the textile industry is to recover and expand markets and employment.

TEXTILES IN NIGERIA: THE SETTING[1]

The size of Nigeria's population is not known. Figures in international reference works suggests some 100 millions by the mid-1980s (World Bank 1987). These are mark ups of dubious census figures from the early 60s. Margins up and down may be in the order of 10 millions or more. By African standards, Nigerian textile consumption was high already in the pre-colonial period. Cloth was favoured in the long-distance trade with Europe and the Arab world. There was a flourishing domestic crafts industry based on local cotton. During the colonial period, Nigeria experienced a high rate of commercialisation of peasant production, She became a major exporter of cocoa, groundnuts, palm produce and cotton. A peasant-based export economy allowed for a substantial non-farming population, which in turn created openings for commercial food production. Peasant and non-peasant consumers alike continued to give priority to textiles, not just for basic clothing, but for a wide range of ceremonial purposes. Personal wealth and social status, even more than in industrial societies, were reflected in the quantity and quality of cloth stored and occasionally displayed.

Local cottage production was swamped by imports, in the classical manner of colonial deindustrialisation, although surviving in niches. The cloth trade was the backbone of the colonial merchant firms. No modern local textile industry was allowed to develop until these firms -fearing Independence - much belatedly spearheaded the establishment of mills aimed at protecting their own markets. This was in the late 1950s. After Independence, textiles rapidly became the leading sector for import-substituting industrialisation. Protectionist policies encouraged by the Civil War (1967-70) brought about a spate of fresh investments. The expansion of incomes from petroleum exports in the post-Civil War era sustained accelerated investments and diversification into more advanced techniques and products. By 1980, with some 100 major plants and some 100,000 workers, the large-scale Nigerian textile industry was important by world standards. In Africa, only Egypt and South Africa could claim larger production capacity (ITMF 1984). In the Nigerian context the textile industry was a leading sector, with about 20 per cent of employment and 15 per cent of value added in manufacturing (FGN 1981). A multitude of small informal producers unrecorded in official statistics, would have to be added. (They are however more numerous in the garments sector which lies outside the scope of this study.)

Oil-generated public and private spending led to an overheated economy, high inflation, high imports and sectoral dislocations, including a heavy outflow of labour from agriculture. This process reached its peak during the years of civilian government, 1979-83 (the Shagari regime). The textile industry suffered from sharp competition from imports, mostly smuggled, high costs, and an overvalued exchange rate. A period of stagnation therefore set in already before the collapse of the oil market in 1981-82 that hit the Nigerian economy as a whole. The foreign payments crisis was temporarily postponed but ulitmately aggravated by indiscriminate foreign borrowing. Brakes were pulled on imports in April 1982 only at a point when commercial credits threatened to become unmanageable. By mid-1983 Nigeria's creditors imposed a virtual embargo on new loans.

By the end of 1983, the civilian regime was ousted by a military junta (the Buhari/Idiagbon regime, 1983-85, the Babangida regime, 1985-). The military offered to reestablish international credit-worthiness. By this time the manufacturing sector had been starved of essential imports for over a year. The new regime reinforced austerity measures, including sharp cuts in public sector spending and employment. Negotiations with the International Monetary Fund (IMF) which had begun under Shagari ground to a halt as no agreement was reached on the size of devaluation, the domestic price of petrol, and some other adjustment measures. The junta, however, pushed ahead with sharply recessionary policies, without the IMF loan.

The Nigerian currency, the naira, which for long had been kept close to *at par* with the US dollar was allowed to slide. But it was only in 1986, as the economy had been hit by a further fall in oil prices, that a substantial devaluation was enforced with the technical and financial assistance of the World Bank. A system of weekly foreign exchange auctions (SFEM) was introduced, raising the price of the dollar some three to four times.

In the case of the textile industry, the crisis predated the general recession. Sharp cuts in employment were experienced already in 1980-81. Firms had embarked on their own "structural adjustment" programmes. The industry was therefore in a better position than others to meet the recession, especially as the foreign exchange crisis also hit at the illegal textile imports which had been a major worry. There was recovery within a market which had drastically shrunk and with a much reduced labour force.

This was the point when the shortage of raw materials emerged as the all-overshadowing threat to the survival of the industry. Domestic cotton production had been allowed to lapse under the impact of the oil boom. Firms opted for cheap and unrestricted imports. The growing importance of man-made fibres had made the industry even more vulnerable to the import squeeze. While continuing to struggle over the allocation of foreign exchange, the textile firms, just like other manufacturers, were forced to look inward, to the local sourcing of inputs. The stage was set for industry to go farming.

I. A New Phase in the Relation between Industry and Agriculture

PEASANTS OR PLANTATIONS: A CLOSING CIRCLE?

In 1921, Sir Hector Duff, a representative of the Empire Cotton Growing Corporation and thus the Lancashire textile manufacturers, toured Nigeria to inspect a troubled cotton industry (Duff 1921). For some time there had been strong lobbying by British industrialists for the opening up of the West African colonies to plantation agriculture (Hopkins 1973). Why should they be restricted when their Belgian, French and German competitors were given a free hand? If East and Central Africa could be opened up for British enterprise, why not West Africa?

Sir Hector was against plantations for two sets of reasons: political and economic. Plantations, he argued, would inevitably lead to conflicts between white landlords and employers and African tenants and labourers. "More than 50 per cent of the problems with which our various East African Governments have to deal", he claimed, took their origin in these contradictions (Duff 1921:62). It was not because the Africans were "unfairly treated", he said, but because the conflicts were inescapable: they were built into the very relations themselves. For this reason, he thought, one should avoid plantations.

This was the more the case as they were unnecessary from an economic point of view; independent African cultivators could serve long-term imperial interests more effectively. Plantations would face difficulties in recruiting, retaining and disciplining labour.

> It was not possible for a comparative handful of Europeans, dependent on hired labour ... to develop the almost boundless resources of tropical Africa as quickly, fully and cheaply as can be done by enlisting the activities of the entire native community on a basis of independent cultivation (Duff 1921:64).

The history of cotton growing in Africa, according to Duff, vindicated this position. The "peasant line", of the imperial ruling class prevailed over the plantation line. Political alliances had already been struck with African land-controlling classes ("indirect rule"), and there was already a history of organised resistance to colonial encroachment of traditional land rights which had to be considered.

In 1985, the year of the Independence Silver Jubilee, the bourgeoisie of Independent Nigeria was shedding the self-imposed restrictions of the former imperial bourgeoisie. The leap into plantation agriculture was well on the way. Mr. H.B. Chatusvedi of the President Clothing Company in Lagos announced that his company had acquired 2,000 ha in Oyo State for experimental cotton cultivation. Total investment costs were estimated at 15 million naira (Business Times 27/5 1985). Mr. Govind Chanrai, the Managing Director of Afprint, a major Lagos firm, informed shareholders that the company had reached an agreement with one of the River Basin Authorities for the establishment of a 2,200 ha cotton project

(Financial Punch 25/9 1985). The General Manager of the Nigerian Textile Mills (NTM), Mr. Oludotun Ilo, claimed that NTM had committed 23 million naira to a 20,000 ha cotton plantation in Gongola State (Guardian 24/10 1985). The Nigerian Textile Manufacturers Association (NTMA) had plans of its own to go into large-scale mechanised cotton production, according to its vice-chairman, Alhaji Mahmud Abubakar (New Nigerian 7/7 1986).

Why is it that the industrialists of independent Nigeria are going into plantation agriculture? Why is it that the peasant farmers who served Manchester so well are no longer trusted by Lagos, Kano and Kaduna? This study is about an important new phase in the development of industry and agriculture in Nigeria and in particular in the relationship between the two. Industry is moving into agriculture on a large scale, the textile industry being only one among many to do so. It is a process with far-reaching repercussions for the development of social relations of production, class formation and for the Nigerian political economy as a whole. It is not just a question of plantations. A variety of methods are employed to subordinate agricultural land and labour to the requirements of industrial capital.

The process is gradual and long-term. A qualitative leap, however, has been precipitated by the current import squeeze and raw material crisis, which is part of the general crisis of Nigeria's import-oriented model of development. As all crises, it provides an arena for intense struggles among social forces, seeking to protect their interests, or taking advantage of the crisis for the purpose of advancing positions.

IMPORT SQUEEZE AND INDUSTRIAL CRISIS

In April 1982, the Nigerian government responded, much belatedly, to the dramatic decline in oil earnings. An "Economic Stabilisation" programme was announced, including restrictions on imports. Chaotic and uncoordinated short-term borrowing and the non-payment of trade bills sustained some imports for another year. By mid-1983, sources of credit dried up. It became evident that Nigeria's Central Bank had lost control over the foreign payments situation. As a result, Nigeria was threatened by a virtual trade and credit embargo.

Chief J.O. Udoji, the President of the Manufacturers' Association (MAN), lamented: Our factories are closing down because of the shortage of raw materials and spare parts; workers are being laid off; commercial banks refuse orders for raw materials and overseas banks refuse to confirm letters of credit. "Our image in the international scene has never been so low" (NISER 1983a).

The reestablishemnt of Nigeria's international credit-worthiness was indeed a major objective of the army officers who seized power from the discredited Shagari regime at the tail end of 1983. By mid 1986, however, the situation had further deteriorated with the new slump in oil prices. A new dangerous experiment in foreign exchange management was launched with the introduction of a second-tier foreign exchange market (SFEM - see p 47. below). To the Nigerian industrial working class the crisis was of disaster proportions. Current levell of imports were capable of sustaining less than half of "normal" capacity. Many firms closed down altogether, others retained a skeleton staff. Workers were sent on compulsory leave with reduced pay or no pay at all. All firms reduced their labour force, also those which were more successful in the access to import licences. The workers who remained faced great uncertainties. Already by the end of 1985, real wages had been cut by half as compared with peak 1981 levels due to sharp price increases. Wages were further badly hit by the massive devaluation of 1986. Trade unions joined the manufacturers in appealing to the state to release foreign exchange for the importation of raw materials and other essentials for industrial recovery.

INDUSTRY AND UNDERDEVELOPMENT

This report is about the structural changes imposed on Nigerian industry as a result of the import squeeze. It deals with the manner in which industry is forced to look inside for raw materials in a situation when outside supplies are cut. There is an intense new public concern in the media over "local sourcing" and "backward integration". Government spokesmen exhort industrialists to shed their dependence on imports. Particular attention has been directed towards those industries that have a potential raw material base in agriculture such as the textile industry.

The weakness of the sectoral linkages between industry and agriculture features prominently in the standard descriptions of underdeveloped economies. One of the major contributions of underdevelopment theory has been to highlight the "disarticulated" structure of these economies and its roots in the way they are integrated in a subordinate position in the world capitalist economy (Amin 1974, 1976; Ake 1981). Each sector has been incorporated on the world scale in a way that obstructs inter-sectoral linkages locally. There is ample evidence to support the "development of underdevelopment" argument in this field; an active entrenchment, pushed from outside, of such disarticulated economic structures.

The products and processes of such underdeveloping industries are so designed as to make them dependent on inputs provided from outside, often from sources controlled within vertically integrated transnational firms. Brand names, patents and product design, coupled with state protection and agressive marketing make third world consumers rely on products with this built-in dependence on imports. Langdon (1981) has demonstrated this effectively in the Kenyan case. Nigerian "agro-based" industry offers good examples.

NIGERIA'S UNDERDEVELOPED AGRO-INDUSTRY

Agro-industries in Nigeria have been heavily dependent on imported raw materials. Among themselves they have been responsible for some of the largest import items, including wheat (flour mills), vegetable oils (oil mills, detergent factories), maize (feed mills), barley (breweries), milk powder (dairies, milk reconstitution factories), sugar (cubing plants), rubber (tyre manufacturers), and of course, cotton and other fibres for the textile industry.

The flour mills are a particularly notorious case. As we have discussed elsewhere (Andrae and Beckman 1985), a two million tonnes milling capacity was permitted to establish itself, based on wheat, an imported temperate crop which cannot be grown economically in Nigeria in any significant quantities. The wheat processing technology is exclusively geared to wheat. It cannot, for instance, be used to process maize. The importance of ownership and transnational links for the entrenchment of such an inappropriate industry is glaring. A major part of the milling capacity was controlled by a US-based firm, a shipping company which handled the massive flow of US wheat. In 1986, the milling and bakery industries were thrown into turmoil by a government decision to ban the further importation of wheat as from 1987.

The breweries, another major industry which has miraculously flourished, crisis or no crisis, were using imported barley, another crop unsuitable for domestic cultivation. Until recently, brewers insisted that only barley will do if proper beer is to be brewed. Most of them have had their formulae tied to foreign patents and brand names controlled by their technical partners.

The dairies is another industry that has been allowed to establish itself almost exclusively on the basis of imported raw materials, despite continuing lip service to the need to harness the vast resources of the Nigerian livestock industry. What is featured as dairies in Nigeria are nothing but simple reconstitution lines for imported milk powder that has been dumped in the world market with heavy government subsidies by mostly West European surplus producers (Andrae and Beckman 1982).

In the case of sugar, an even bigger drain on Nigeria's foreign exchange, some local production has been developed (at the most 20 per cent of consumption), and some more is in the pipe-line, although the problems confronting the domestic sugar manufacturers seem formidable, as demonstrated by the half a billion naira Savanna Sugar Company in Numan (Andrae and Beckman 1982). No wonder that Tate and Lyle and other "manufacturers" until recently resisted involvement on the agricultural side, preferring the easy way of importing granulated sugar for either cubing or mere packaging.

More surprising has been the massive imports of maize for the booming feed mills, almost exclusively linked to the poultry industry. This is the more remarkable as maize grows well in Nigeria and production has experienced a high rate of increace in the past decade. Here again, the transnational links are important. A US-based firm, Pfizer, dominates the industry. In the wake of a bumper maize crop in 1985 and in the face of the deepening foreign exchange crisis, the government finally decided to stop further maize imports.

TEXTILES: THE RAW MATERIAL THAT DISAPPEARED

The textile industry is different. Although it is currently a big importer of raw materials this was not always so. In 1973, the output of Nigeria's cotton farmers was roughly equivalent to the requirements of the industry. Ten years later, domestic production had fallen to less than 40 per cent of the cotton used by the industry and half the quantity of 1973.

Textiles had been Nigeria's leading industry with some 100 factories and some 100,000 workers, leaving out the uncounted numbers engaged in small weaving establishments, including those in the informal sector. In line with long-established theories it has, as a producer of essential mass consumer goods, it has become the backbone of early industrialisation. Its attraction is enhanced by its use of relatively labour intensive and widely available, standardised technology. Its ability to draw on an already existing raw material base made the textile industry even more attractive as a centerpiece of national development.

Cotton was an important export crop, the production of which expanded fast in the post-second world war period. It was therefore merely a matter of redirecting production from exports to domestic use. An elaborate net-work of state marketing, extension services and distribution of inputs was already available, as well as an intermediate processing industry, the ginneries, a dozen of which were dispersed over the cotton growing areas of northern Nigeria, from Sokoto in the west to the Upper Benue Valley in the east.

Why did cotton production decline so drastically? This is discussed in the third part of the report. The case of the textile industry, however, demonstrates that the raw material crisis of Nigerian agro-industry goes beyond the entrenchment of inappropriate and import-dependent production. The crisis of the textile industry is as much a crisis of the domestic agrarian base, althouth exacerbated by the import squeeze. Dependence on imports had grown in step with the decline in domestic cotton supplies. As we shall argue below, imports themselves contributed to this decline.

VEGETABLE OILS: MORE DISAPPEARING RAW MATERIALS

The textile industry is not unique in facing a disappearing domestic raw material base. Nigeria was once one of the world's leading exporters of palm produce and groundnuts. Beginning in the late colonial period, processing industry was established based on these crops, primarily intended for exports but also for a fast growing domestic detergents industry (Kilby 1969). Since the Civil War in the case of palm oil, and since the early 70s in the case of groundnuts, the vegetable oil mills were either closed down or operated on imported raw material. Some of the companies simply abandoned processing and turned to trading, sometimes pathetically exhibiting their imported produce as "Made in Nigeria" after putting it on tins locally.[2]

In the case of palm oil, however, agro-industry continued to flourish at a lower level. The big mills were not able to compete for the produce with thousands of small cottage-type refineries. Total output may have declined, but above all, produce was redirected into the expanding domestic market. In the case of groundnuts, production declined more drastically, especially as land and labour were redirected into commercial grains production for local food markets. There was also a general movement of labour out of agriculture as a result of the petroleum-fed growth on non-agricultural economic activity.

HAS THE PEASANTRY OUTLIVED ITS USEFULNESS?

The dramatic decline in Nigeria's export crops was taken as evidence of the crisis of the peasant economy. Together with the massive increase in food imports it gave support to arguments for the restructuring of agricultural production on capitalist lines. The peasant economy was seen as outmoded and redundant, unable to accommodate the necessary technological changes.

For the past decade such assumptions have been allowed to justify an accelerated shift in Nigerian agricultural policies in support of large investments either aiming at bypassing the peasantry altogether or at the restructuring and subordination of peasant land and labour through contractual arrangements and restrictions on peasant control of land. More importantly still, such assumptions have legitimised political and financial support by the state for a growing class of capitalist farmers. Many link up with foreign technical partners who are actively enticed by the state to participate in "revolutionising" Nigeria's agriculture. Restrictions have been removed on direct foreign investment in this sector. Ambiguities in existing land legislation, purportedly aimed at protecting the interests of the "weak", are exploited by the authorities for the appropriation of land on behalf of such new commercial investors, domestic and foreign. (Beckman 1987a; Oculi 1984).

We speak of the assumptions about the redundant peasantry as justifications because the real motive force behind this transformaton must be sought in the changing balance of forces and the rise of capital, partly compelled, partly anxious to extend its frontiers of accumulation into agriculture. The current crisis provides excellent conjunctures for these forces to advance their positions. The direct involvement of industrial capital is an important new development.

INDUSTRY GOES FARMING

A new fever spreads through Nigerian industry: Farm or die! Since 1985 company annual reports are full of references to farm projects, sometimes directly related to the procurement of raw materials, sometimes more as a means of diversification of the enterprise. The Nigerian Enterprise Promotion Board reports on a "tremendous upsurge" in the interest shown by foreign firms in agricultural projects (Financial Punch 7/8 1985). On the occasion of the Silver Jubilee of Independence in October 1985, newspapers featured congratulatory advertisements from companies anxious to demonstrate their patriotism by going into agriculture. Food Specialties of Nigeria, the importer of powdered milk (a subsidiary of the giant food multinational Nestlé) announced a first phase of a large-scale farming project. The company had incorporated a 2 million naira subsidiary and applied for 2,000 ha of land. Journalists were taken around a pilot farm by international experts (National Concord 11/10 1985; Guardian 25/7 1985; Business Concord 2/8 1985; New Nigerian 15/8 1985).

Soft-drinks producer 7-Up, in its own Silver Jubilee contribution, recognised "the urgent need to help to relieve the financial burden of the national economy". Its 10-12,000 ha fully integrated sugar project would save US$ 150-250 in foreign exchange, it claimed (New Nigerian 17/10 1985). John Holt, Nigeria's number two trading and industrial conglomerate, a subsidiary of Lonrho, was also keen to project a new image of serious concern with agriculture. A big private farm had been bought (Ogbemudia Farms), and a 10 million naira investment was planned, including an oil palm plantation. "Having participated actively in every aspect of

Nigeria's commercial and industrial growth John Holt is poised to make its contribution to the Agricultural Renaissance" (Guardian 11/10 1985).

While some of this was public relations exercise of little consequence, the general thrust is not to be mistaken. The political pressures are also real. The breweries which not so long ago spoke of the impracticability of local sourcing (cf. Business Times 4/6 1984), were the first to be confronted with a government time table. In a policy statement to the "stunned Chief Executives of Nigeria's brewing firms hurriedly summoned to Lagos", the Federal Minister of Commerce and Industries declared that they were given ten years to achieve 100 per cent local content. Already by 1986, 25 per cent of the imported barley should have been replaced with local grain (Business Concord 5/4 1985). Some of the leading brewers soon announced their farming plans, including N.B.L., "Nigeria's most profitable industry", which had acquired 15,000 ha in Niger State (Business Concord 7/6 1985; Guardian 17/6 1985, 27/6 1985).[3]

Of particular historical significance, was the announcement by Lever Brothers, Nigeria's leading soap manufacturers, a subsidiary of the United Africa Company, Africa's number one multinational, that it would establish a 100 million naira oil palm plantation (New Nigerian 1/6 1985). It may be recalled that it was the attempt by the predecessors of this company to obtain land for oil palm plantations which sparked off the political controversy in the early colonial period referred to in the introduction above. A closing circle.

FARMING FOR IMPORT LICENCE

Faced with the import squeeze and mounting government pressures some industries went into farming, not so much in order to produce their own raw materials but to demonstrate their good will to the government, thereby qualifying for import licences and foreign exchange allocations. One brewery stated this explicitly: The growing of local food crops should entitle it to an equivalent amount of licence to import barley (Business Times 23/4 1984, 3/9 1984). This was also at first the strategy of the Flour Mills which had no intentions of replacing the imported wheat (cf. Guardian 29/10 1985). It was only when the government banned the importation of wheat (to be effective from 1987) that maize milling and maize farming became a serious concern to the wheat millers. Also firms which did not use imported agricultural inputs resorted to such a strategy. The chairman of BEWAC, Chief Adewale, was quite blunt on this point. His company had decided to go into farming because that had "become a pre-requisite for patronage by government in the allocation of import licence" (Business Times 23/9 1985). In opening the fifth Lagos International Trade Fair in November 1985, President Babangida made the conditions explicit: The allocation of import licenses would be premised on the ability of an applicant company to conserve foreign exchange through the use of local raw material (Triumph 4/11 1985). Clearance would be required from the relevant commodity boards (New Nigerian 18/11 1985).

There were other good reasons for companies, especially foreign-owned ones, to seek to please the government in this respect, including the need to facilitate the transfer of profits and the employment of foreign personnel, all affected by the foreign exchange squeeze. In the case of the petroleum multinationals, there was special pressure to make themselves useful by reinvesting some profits in agriculture. This was considered particularly appropriate as the coming of petroleum was seen as a cause of agricultural decline. As petroleum was the villain, it should also make up for it, somehow. Several of the oil companies have gone into farming including Total which acquired a 6,000 ha farm in Oyo State (National Concord 16/8 1985).

The enthusiasm and seriousness with which industry was entering into farming varied. Some of it was window-dressing (See the "Executive Suite" interview with Longe in Financial Punch 13/11 1985). All sorts of excuses for stalling were found, including government's failure to secure land on reasonable terms or to allocate import licences for agriultural machinery, chemicals, foreign farm managers etc. Government policy guide-lines were easily subverted, as they had been in the past, through private deals and special considerations. Government guide-lines themselves were often unrealistic, inconsistent and not seriously pursued by the authorities. Yet the pressure to integrate backward into agriculture was part of the logic of capital accumulation. Investments had been made in industrial capacity which could no longer be utilised fully. Capital was being wasted. The owners and managers of capital were seeking solutions for cutting the waste.

II. Textile Industry and the Raw Material Crisis

SIGNS OF CRISIS

The major outcry over the raw material shortage in the textile industry came only in mid-1984. A year earlier the chairman of the Nigerian Textile Manufacturers Association, NTMA, Mr Oludotun Ilo, was quoted in the press warning of the uncertain times ahead for the industry as a consequence of the large gap of tonnes in the expected allocations of import licences for cotton (Business Times 29/3 1983).

The pitch of the outcry at the time of licence allocations in 1984, gives evidence that the full impact was now being felt of the shortage situation. The textile industries, individually and through their organisation, NTMA, were joined by the unions of senior and junior employees in pointing out the ill effects of the shortage and painting the picture of doom if import licence allocations were not raised. Labour redundancies, low capacity utlitisation and even the close-down of industries were part of the picture painted.

Other reports in the press speak of low capacity utilisation in a number of factories. For example, Five Star, a major lace and knitted goods producer in Lagos was running below 50 per cent of its daily capacity of 100,000 metres, working only 4 hours per day, and threatening to lay off 670 workers of its about 1,000 workers. It had received only 3 million naira worth of import licences, while applying for 18 million. The previous year the problems had been mitigated by remaining stocks from 1982; now there was no more buffer (Guardian 1/7 1984).

In the words of the General Secretary of the National Union of Textile Garment and Tailoring Workers of Nigeria, Mr Adams Oshiomhole, 8,000 workers had been made redundant since the beginning of the same year. Of the remaining workers at least 6,000 in various mills were said to be working only 4 days a week. Many had been sent on compulsory leave. All resulting from the fact that the low import allocation had allowed the industry to operate at only 40 per cent of its capacity. With the current allocation assumed to cover only 30 per cent of requirements, an SOS had been sent to the military government imploring it to give special consideration to the labour intensive industries in the allocation exercise (New Nigerian 25/6 1984). On a similiar note the president of the Textile, Garment and Tailoring Senior Staff Association, Chief Chinyam Ubur, maintained that 18 factories had been closed down and warned of the loss of another 20,000 jobs before the end of the year (Business Times 16/7 1984).

This picture is confirmed by later compiled overall data: In a survey of our own we record a drop in capacity utilisation from 1983 to commonly below 50 per cent in 1984: Of 11 companies for which data are available for 1984 or 1985, capacity utilisation is found to lie between 30 and 50 per cent in 8 of them, while two had 60 and 75 and one was down to 9 per cent. We find a drop in labour utilisation, above all by stoppages in production of one to two months with compulsory leave for workers, when raw materials run short. Further we find a downward trend in employment figures which also accelerates in 1983-84: Of 10 surveyed factories 9 were found to have lost altogether 3,832 workers, or 20 per cent, between 1982 and 1984. The Nigerian Textile, Garments and Tayloring Employers' Association find that of 47 recorded establishments 25 per cent employ less than 50 per cent of their full labour force in 1984 (interview Eburajolo 15/11 1985).

Thus there is crisis no doubt, hitting some companies more than others, as usual, but reaching all in some measure: The index of production compiled by the Central Bank of Nigeria, where 1972=100, went from 258.5 in 1982 to 103.9 in 1984 (and 93.5 in 1985) for cotton textiles, and from 1,175.0 in 1982 to 639.4 in 1984 (and 560.6 in 1985) for synthetic fabrics (CBN Monthly Report, various years). Let us look closer at the size and structure of the production apparatus affected.

THE TEXTILE INDUSTRY: INSTALLED CAPACITY

The large-scale industry in Nigeria is organised in some 100 establishments with a capacity to produce 400-600 million metres of cloth. The products range from grey baft to shirting, dyed cloths, prints of various quality and pattern as well as embroidered lace. They also include a smaller share of towels, bedsheets, blankets and carpets as well as some knitted goods. 67 of the textile factories were members of the Nigerian Textile Manufacturers Association in 1985 (together with a few garment factories, that are not included in this study). For these registered factories data were available referring to mid-1985, which we may use to bring out some characteristic features of the industry (NTMA 1985a).

The Association estimates that they cover all but a few per cent of the capacity in the large-scale textile sector by these data (Young Itiye 13/9; 12/11 1985). Their aggregated capacity figures correspond to 800 million meters of cloth, which must be understood as an overstatement. They were offered in the context of application for import licences. The material can be seen to include a few firms that had not been producing for sometime and some capacity that was not immediatly operational according to our cross-checks in the field. Such entries have been deleted by us, making the total number of analysed establishments 64. Relative numbers on capacity may be of interest to analyse from these hard-to-come-by data. Notations on location and nationality were given by the secretary of the NTMA and cross-checked with the secretary of the Textile, Garments and Tayloring Employers' Association (interview Eburajolo 15/11 1985). Nationality designation refers to foreign owner participation, or in a few cases, possibly rather management control. Our source will have characterised as Chinese or Indian some firms which are British registered by such ethnics. Chinese in these cases means Hongkong Chinese. The following features of the industry emerge from this material, as summarised in Tables 1 and 2.

26 establishments account for all the spinning and are thus the ones to compete for the domestically available and imported cotton fibre. Most of these factories are large, operating from 25,000 to some 55,000 spindles. Only a few have less than

Table 1 *Textile factories: Location and nationality (NTMA members)*

Spinners[a)]	Indian	Chinese	Lebanese	Nigerian	Other	Total
Lagos	3	2	1	1	1	8
Kano			4	1	1	6
Kaduna		2		1	1	4
Other South		1		3		4
Other North		3		1		4
Subtotal	3	8	5	7	3	26
Non-Spinners						
Lagos	24	1	1	3	1	30
Kano	1		1	1		3
Kaduna	2	1				3
Other South				2		2
Other North						0
Subtotal	27	2	2	6	1	38
Total	30	10	7	13	4	64

Note: a) Those who spin at all. (All but 5 also weave and/or knit).
Source: NTMA 1985a.

10,000 spindles, one has over 80,000. Most of these spinners also weave and together they account for some 80 per cent of available weaving capacity. Two thirds of the spinners are located outside Lagos, four in Kaduna, including three of the largest ones, six in Kano, and one each in Gusau, Funtua, Ilorin in the northern states and in Aba, Ado-Ekiti, Onitsha and Asaba in the South. Our classification by nationality shows that notable among owners of these large integrated establishments is a number designated as Chinese. Five of these are affiliated with the Cha group which has its base in Kaduna. This formally British-registered firm dominates a group of eight Chinese controlled factories. A similar number of factories are designated as Nigerian, spread out over the country. Five spinners are Lebanese, four of which are in Kano. Three have Indian participation, all Lagos-based.

In addition to the 26 spinning establishments, another 17 are combined weavers, knitters or embroiderers. These are centered in Lagos, with a few also located in

Kano. Some 18 of the non-spinners were single-function firms. They were weavers in most cases, usually smaller ones. Some were found in Kano; most of them in Lagos. Printing was done as part of the finishing process by an indeterminate number of weavers, and occasionally as a single activity. Of the altogether 40 non-spinning firms, 27 had Indian participation, all but three of which were located in Lagos. Apart from six Nigerian-controlled factories also centred in Lagos, other nationalities were few of a kind in this category. This information is summarised in table 1.

Apart from the NTMA member firms, another 15 smaller ones were known to the Association. Their emphasis was in knitting (about 15 per cent of total recorded capacity) and in embroidery (about 7 per cent of such capacity). Nationality in this group was only sketchily known.

The strong foreign dominance in the industry can be summarised by aggregating the NTMA figures on capacity, as in the first half of Table 2. By decree at least 40 per cent of the capital in every spinning firm, and 60 per cent in other textile firms was Nigerian. But is was only about one third of the capacity in conventional spindles and looms that was found in factories that were designated as all-Nigerian in our material. In the more modern rotor spinning and shuttleless weaving as well as in knitting this share was much smaller still. This was rather where Indians feature strongly, importantly in shuttleless weaving, to a lesser degree, and together with the Lebanese-origin owners, in rotor spinning, and overwhelmingly in knitting and embroidery. In terms of volume, the conventional spinning and weaving processes, however, constituted by far the dominant part of the sector, and here the Chinese represented the main foreign element. A technological bias according to nationality is thus another feature in the industry. The predominance of Kaduna as a centre for conventional spinning and weaving stands out from the same data, as arranged in the latter part of Table 2. Kano appears as more modern-machinery oriented in spinning and weaving, and has some embroidery as well. Lagos is shown to have the most complex all-round structure.

Product orientation is however the crucial variable in the present context. Whether fibre or yarn is demanded and whether these inputs are sourced in imports or in domestic supplies has become a matter of life and death for the individual firms in the current raw material crisis. Technological change becomes an important factor when it comes to strategies of adjustment to the crisis.

TRENDS TOWARD RAW MATERIAL SHORTAGE

How could the import squeeze so badly affect an industry that had originally developed on the basis of large domestic supplies of cotton fibre? We have said above that in Nigeria the raw material gap in many agro-based industries is the result of an expansion with little regard for domestic supply potentials, but that initially, from the emergence of the textile industry in the 1950's the situation was very different in this sector: A long-established domestic supply of cotton that was geared to export, was gradually expanded and redirected for absorption in an industry that was in fast expansion.

Table 2 *Textile industry: Type of process by nationality and location. (NTMA members.)*
Percentage of installed machinery.

	Spindles	Rotors	Shuttle looms	Shuttleless looms	Knitting	Embroidery
Nationality						
Indian	13	31	17	63	86	88
Chinese	36	7	46	0	3	12
Lebanese	6	36	0	13	0	0
Nigerian	34	19	28	11	11	0
Other	11	6	10	13	0	0
Total	100	100	100	100	100	100
Location						
Lagos	33	40	39	69	97	60
Kano	9	42	3	22	0	7
Kaduna	33	18	33	8	0	0
Other South	15	0	16	1	3	0
Other North	9	0	8	0	0	0
Total	100	100	100	100	100	100
Number of machines	662,268	4,893	18,409	2060	811	215

Source: NTMA 1985

This trend was, however, broken, when a surplus in domestic fibre developed and combined with a preference for imported cotton yarns leading to a fall in the demand for domestic fibre. This fall contributed to a drastic decline in fibre supply. Favourable conditions for imports at this time made it possible to maintain raw material supplies at a peak volume in spite of receding domestic supplies. Import dependence both for yarns and fibre thus developed to a point were the import squeeze in 1982 was bound to severely affect production potentials from then onwards (or as soon as stocks were depleted). The effect was particularly severe as it was added to a growing dependence on imported man-made materials. By 1982 the textile industry was no more very different from other agro-industries in terms of import dependence.

Table 3: *Supply of raw materials to the textile industry.*

Year	Cotton materials						Man-made materials[a]				Total supply
	Fibre Domestic less export		Imported		Yarn Imported		Fibre Imported		Yarn Imported		
	tonnes '000	%	tonnes '000	%	tonnes '000	%	tonnes '000	%	tonnes '000	%	tonnes '000
1973	39,8	64	0,03	0	2,2	4	1,8	3	18,8	29	61,8
1974	20,3	30	17,2	25	1,2	2	10,6	16	18,2	27	67,5
1975	51,2	46	15,8	14	4,4	4	6,3	6	34,9	30	112,6
1976	58,4	52	2,0	1	6,9	6	4,2	4	40,4	36	111,9
1977	71,0	55	0,2	0	0,3	0	3,4	3	57,4	47	128,3
1978	30,0	22	1,3	1	23,0	17	2,6	2	75,0	54	139,0
1979	36,4	30	0,2	0	20,1	16	2,5	2	63,2	52	122,4
1980	28,5		n.a.	n.a.		n.a.		n.a.		n.a.	
1981	26,4	21	15,7	13	19,4	16	1,8	1	60,5	49	123,8
1982	21,6	26	23,0	28	10,1	12	4,7	6	22,6	28	82,0
1983	19,9	20	25,4	25	8,6	9	5,2	5	38,3	38	99,9
1984	12,6	14	25,7	29	4,9	6	5,0[b]	6	41,2	46	89,4

Notes: [a] The figures for man-made materials include small amounts of other yarns, e.g., wool and flax. Chips for production of man-made figres are not included. [b] N.a.: Assumed figure.
Source: Cotton production data: CAP (1985b); refer to crop year ending specified year. Import data: FOS, Nigerian Trade Summary; refer to calendar year specified.

These parallel developments appear from the compilation of official import data and figures on domestic cotton supplies from the ginners. They are presented in Table 3 and Diagramme 1, on which we may further elaborate as follows: When in the 1970s the total raw material supply was raised from 60,000 tonnes per annum to 110,.000 in 1976 it still contained a component of close to 50 per cent domestic cotton fibre. This proportion was kept up and even somewhat expanded in the following two years of continued total supply expansion. In addition, some limited exports continued up to 1979. After a few years of thus increased total raw material supply with a growing local base, there was in 1978 an abrupt fall in the domestic fibre supply, from 70,000 to 30,000 tonnes per annum. This initiates a period characterised by a completely different supply situation, dominated by yarn imports. The increase in supply above 50,000 tonnes of cotton fibre in the 1975-77 period will have exceeded domestic spinning capacity by some 10-20 000

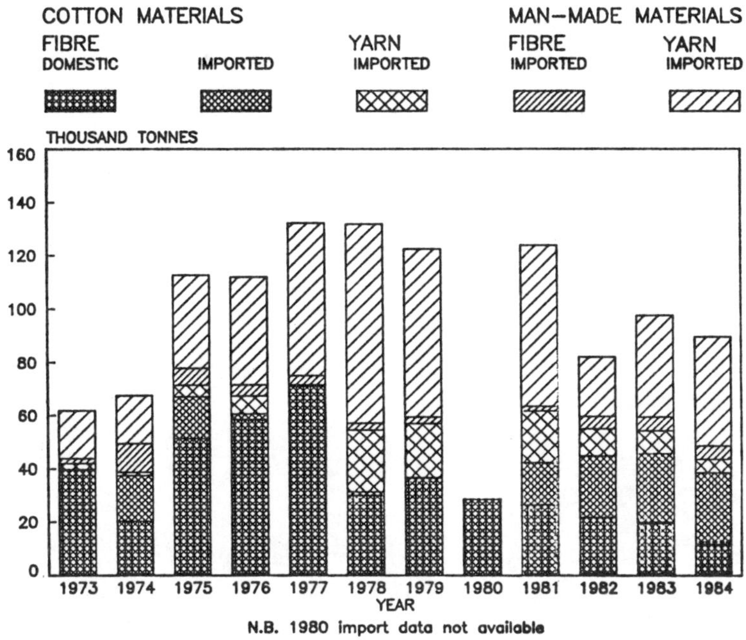

Diagramme 1. *Supply of raw material to the textile industry.*
Notes and sources same as in Table 3

tonnes per annum. This implies that a surplus would have been available to the spinners in the following two to three years at least. The cotton board officials tell of years when millers had to be implored and begged to buy up the surplus in advance of their needs, in order to provide storage for the bumper harvests of consecutive yers. It may be assumed that this will have weakened the pressure on domestic cotton production and it contributes to the explanation why little was done to reverse the decline when it set in.

As will be further discussed below, several processes at the agricultural end will have contributed to this decline in domestic cotton production. A further factor that can be ascribed to decreasing demand from the industry is suggested by our present data: In the same year as the domestic cotton fibre supplies drop in 1978, there appears for the first time a large item of some 20,000 tonnes of imported cotton yarns that remains part of the picture of material supplies up to 1981. Possibly these cotton yarn imports were initiated in response to the needs to supplement the receding cotton fibre supplies, a need that would have been felt in spite of the surplus from earlier years (the surplus will for instance have been unevenly distributed between industrial firms).

Very likely this factor will have been reinforced by the attraction of favourable prices of imported yarns as compared to the locally spun varieties, of which millers report in our interviews later in the 1980s. A sharp increase in the price of Nigerian cotton fibre since the 74/75 season (to be discussed below), will have contributed to this relative price attraction of the imports. Apart from this, a changing product orientation, requiring a broader variety of yarns, that could not readily be supplied in Nigeria, will have supported the process.This will in turn have been enhanced by the competition from smuggled goods that was building up at this time. The receding domestic cotton supplies were no threat to the industry in this situation particularly not as importation of man-made yarns as allowed to increase by the same large amount. From holding a share of one third of total supplies up to 1976 it increased to around 50 per cent of total supplies, or more than 60,000 tonnes per annum.

When the import squeeze struck in 1982 the decline in domestic supply of cotton had reached a level where it accounted for only 21 per cent of all materials for the industry compared to the 55 per cent only five years before. The pinch of the import squeeze was sharpened as this receding tendency continued unabated over the following years, so that by 1985/86 it was down to a mere 10,000 tonnes. Import dependence continued to grow in the face of decreasing potential to import. Production in the industry had to fall.

THE REQUIREMENT FOR LOCAL COTTON

The supply crisis precipitated by the import squeeze threatens the whole range of supplies for textiles, including chemicals and dyes and machinery spare parts and accessories. Near total import dependence makes these inputs contribute some 25 per cent of the total import bill for the industry. The challenge of substituting them locally has been passed to the chemical, petrochemical and machinery industries. It is treated as a rather long-term project (see e.g. Ilo 1983).

When it comes to the fibre and yarns which are our main concern in this paper, the immediate focus has come to center heavily on increasing the domestic supply of raw cotton. The challenge is one of not only replacing receding supplies but of expanding them several times. Three factors converge to put increased pressure on domestic raw cotton supplies. First there is the simple need to replace the imports of cotton fibre which developed to fill the gap of receding local supplies. Secondly, there is the new demand for fibre that develops with the ambition and necessity to replace the previous imports of yarn in the expanding domestic spinning industry. A third factor which is going to put pressure on raw cotton supply is the immediate need to replace the man-made materials, for which no local production capacity is expected to be available until 1992 at the earliest (see e.g. Ikokwu 1985). Their share of inputs is unlikely to be maintainable in the new import situation. Part replacement by domestic cotton may therefore suggest itself. What does this mean for the size of cotton supplies that would need to be produced domestically?

The extent of the total raw material shortfall as defined by the industry went beyond the difference between the 82,000 tonnes available in 1982 and the 125,000 supplied in a number of the preceding years. When the NTMA chairman, Mr O. Ilo (1983), demonstrated the challenge ahead in securing the raw materials required, he rather departed from the quantities that would be needed to supply the industry at, what he assumed to be, its full capacity. Defining the latter as the production of 560 million metres of cloth and assuming 0.3 kg fibre per metre of cloth, he arrived at a total requirement figure of 168,000 tonnes of fibre. Of these he suggests that 134,000 tonnes of cotton is the amount to be domestically provided. These calculations, aimed at estimating the potential savings in foreign exchange from local sourcing, were offered in a paper contributed to the National Workshop on Raw Materials that was organised in 1983 by NISER and the Federal Ministry of Commerce and Industry.

Table 4. *Raw material structure at different levels of total supply. Thousand tonnes*

A. Actual supply 1982: 82,000 tonnes

	Total	Domestic Fibre	Imported Fibre	Yarn	Subtotal
Cotton	54,7	21,6	23,0	10,1	33,1
Man-made[a]	23,7	-	4,7	22,6	27,3
Total	82,0	21,6	27,7	32,7	60,4

B. Ilo's assumption: total required supply 168,000 tonnes (1982 structure of materials).

	Total	Domestic Fibre	Imported Fibre	Yarn	Subtotal
Cotton	100,8	21,7	47,5	31,6	33,1
Man-made[a]	67,2	-	5,0	60,2	67,2[b]
Total	168,0	21,7	52,5	91,8	146,3

C. Total supply. 125,000 tonnes. (1982 structure of materials.)

	Total	Domestic Fibre	Imported Fibre	Yarn	Subtotal
Cotton	75,0	21,7	32,0	21,3	53,3
Man-made[a]	50,0	-	5,0	45,0	50,0
Total	125,0	21,7	37,0	66,4	103,3

Notes: [a] Man-made yarn includes small amounts of other materials, e.g. wool and flax.
[b] Fibre-yarn distribution not specified by Ilo.
Source: Table 1

Following his assumptions that relations between cotton and man-made fibre would be about 60 to 40 per cent, he arrived at a requirement of 100,800 tonnes of cotton, of which only some 22,000 were domestically supplied in 1982. Of a current requirement of 67,000 tonnes of man-made inputs, all of which would have to be imported, he implied that 50 per cent, or 33,000 tonnes, might be replaced by cotton and subsequently also locally sourced.

We have summarised Mr Ilo's calculations on the current structure of requirements in Table 4. In combination with his assumptions on local sourcing potentials, they imply that a total of 113,000 tonnes should be added to the 22,000 tonnes currently produced, if domestic sourcing is to be achieved. This appears from Table 5. Thus we arrive at a total figure of 134,000 tonnes to be locally supplied. This is six times the supply in 1982, or ten times that in 1983 and 1984.

Table 5. *Domestic cotton requirement at different levels of total raw material use. Assuming full substitution of the imported cotton component and substitution by cotton for 50 per cent of the man-made component. Thousand tonnes.*

Total level of material supply	1982 level of domestic supply	Imported cotton component: fibre and yarn	50 per cent of man-made component	Total domestic cotton required
1982 level: 82,0	21,7	33,1	13,7	68,4
Ilo's assumption: 168,0	21,7	79,1	33,6	134,4
1977-81 level: 125,0	21,7	53,3	25,0	100,0

Source: Table 4

For comparison his mode of calculation is applied to the actual 1982 level of production and supply and shown also in Tables 4 and 5. This yields the more modest short-fall of 47,000 tonnes.Ilo's estimate of 134,000 tonnes is no doubt on the high side. Looking at the figures in Table 3 and Diagramme 1 we may question his assumptions on idle capacity. His figures may rather represent assumed potentials for expanded textile production, if all conditions were fulfilled. It appears that there was less than full capacity utilisation in the industry even before the raw material shortage set in. Aging machinery, erratic electricity supplies, raised labour costs and above all the competition from smuggling were factors that had been affecting the industry around the turn of the decade (see e.g. Auta 1982). We also agree that as a point of departure for the long term strategies, to be further treated in section IV below, an interesting figure to consider is the total capacity requirement of materials; not least for the force toward filling the gap that the idle capacity represent. In the shorter run however the realistic capacity level may in our view be assumed to lie rather somewhere round 125,000 tonnes of total material use, to produce about 420 million metres of cloth, the supply level reached in a number of years in the 1977-81 period. Applying again the same structural assumptions as in Ilo's calculation, we arrive at a requirement of 100,000 tonnes of cotton lint to be locally sourced (see again Tables 4 and 5). This is nearly five times the 1982 supply figure.

III. What Went Wrong with Nigeria's Cotton

THE RISE AND FALL OF NIGERIAN COTTON PRODUCTION

In the self-congratulatory annals of the British Cotton Growing Association, the origins of "commercial" cotton production in Nigeria is put to 1902, when the Association was inaugurated to assist in the development of new sources of supply to the mighty Lancashire textile industry (Cowley 1966).What is often forgotten is that cotton was a major household and commercial crop before the colonialists came, serving the neds of a domestic spinning and weaving industry. It supported not only a high level of local textile consumption but also long-distance trade. Cotton based textile production was a general feature of the Nigerian village economy. Major precolonial cities had vast quarters entirely settled by specialised textile producers (Pokrant 1981).

This flourishing domestic industry and trade was seen as a threat by some of the Lancashire millers, competing as they did both for raw materials and in final markets. The Duff Report (1921) argued that nothing could be done to prevent this competition, but that there was no need to worry. The fate of the local textile industry was sealed: "All the world over peasant industries, in the end have gone to the wall before the competition of modern methods" (Duff 1921:72). He also argued that it was thanks to this domestic production that Nigerian farmers had responded so swiftly to the new, colonial export trade.

The exports rose from 360 tonnes in 1903-04 to 3,600 tonnes in 1916. It was an impressive achievement so far, but the momentum was not kept up. Despite occasional peaks such as the 9,000 tonnes of 1934-35, the inter-war period was characterised by sharp fluctuations and stagnation, including bottom marks as the 900 tonnes of 1931-32 (Cowely 1966). For all this period it is likely that production for local consumption exceeded exports. As late as in 1968, A.D. Little reported on current estimates that as much as 14,000 tonnes were still used directly for home or cottage industry. Tiffen (1976) reports from Gombe that the reduction of textile imports during the second world war resulted in a sharp increase in local spinning and weaving.

The big increase in exports began after the war, influenced by the favourable prices that characterised most of the late 40s and the 50s. The development is shown in Table 6, using the figures from the ginneries, thus covering both exports and the growing domestic industrial consumption which began in the late 50s. We find that the output of cotton lint i.e. fibre, rose from an annual average of a mere 6,000 tonnes in the second half of the 40s to 32,000 in the late 50s, 46,000 in the late 60s and to an annual average of 50,000 in the first half of the 1970s (CAP 1985b; c.f. also NCB Annual Reports and MANFU 1985).

Table 6. *Nigerian production 1945/46 to 1984/85. Thousand tonnes of cotton lint.* [4]

1944/1945	-	1964/1965	46,4
1946	6,1	1966	45,4
1947	5,9	1967	52,4
1948	3,3	1968	27,5
1949	8,7	1969	56,7
1949/1950	11,1	1969/1970	91,4
1951	13,9	1971	39,4
1952	20,6	1972	38,4
1953	17,3	1973	48,4
1954	25,5	1974	30,9
1954/1955	33,6	1974/1975	51,2
1956	27,7	1976	58,4
1957	25,6	1977	80,4
1958	43,1	1978	37,1
1959	31,4	1979	36,4
1959/1960	30,4	1979/1980	28,5
1961	52,8	1981	26,4
1962	29,0	1982	21.6
1963	69,2	1983	19,9
1964	43,9	1984	12,6
		1984/1985	15,0

Source: CAP (1985b)

There were sharp year to year variations. On several occasions output either rose or fell by over 50 per cent. Looking at the 60s, we find dramatic increases in the early and late years of the decade, culminating in the all-time peak of 91,000 tonnes in the 1969/70 season. By 1974, the output had been slashed to 31,000 tonnes, rising again to a new peak in the 1976/77 season at 80,000 tonnes.

It is from here on that the decline sets in. While the remaining years of the 70s were only slightly down on the average for the last two decades, the annual output during the first half of the 80s dropped by more than 50 per cent. The 1983/84 crop was considered a real disaster with 13,000 tonnes, the lowest figure since 1950. The 1984/5 crop showed some marginal improvement, while the figures announced for 1985/86, a mere 10,000 tonnes, seemed to confirm the downward trend (New Nigerian 14/5 1986).

What was the problem? Why were the Nigerian peasant producers deserting cotton at the time the crop began to be most desperately needed by Nigeria's crisis-ridden textile industry? Most of the specific issues have been reported over and over again in the annual reports from cotton research programmes and cropping scheme meetings at the Institute for Agricultural Research in Zaria, the long-established centre of research and extension services in the field of cotton[5]. They were also ventilated at the first National Symposium on Cotton Production, held in Zaria in 1982. In 1985 the Governors of the ten northern states set up a committee

to look into the problems of cotton production under the chairmanship of Dr. Shettima Mustapha (New Nigerian 20/3 1985).
The main problem areas identified relate to low productivity, competition from food crops, high labour costs, and inefficient marketing arrangements. Let us discuss them one by one.

THE PROBLEM OF LOW YIELDS ON PEASANT FARMS

Long before the traumatic decline in production of the late 70s and early 80s, Nigeria's cotton experts and international consultants were disturbed by the discrepancy of yields achieved on experimental farms and what was obtained under peasant conditions. Vastly improved yields could be achieved by following recommended practices based on decades of careful research and which had been widely disseminated by the extension services. Recent survey work seems to confirm that farmers are well aware of the content and function of such improved practices (IAR 1983, 1984; Etuk 1985). So what was the problem?

Improved practices mean in particular early planting, early regular and adequate weeding and application of fertilisers and the spraying against crop pests. A Study Group on Cotton of the National Agricultural Development Committee calculated in 1971 that the Nigerian textile industry would require a doubling of the crop by 1985. The Committee believed that this could be achieved with only a light increase in the total area cropped if the improved practices were accepted.

The obstacles, it was argued, were not technical but social and economic (Faulkner 1974; Norman 1974; cf. IAR 1983, 1984). Very little investigation, however, was made into these social and economic problems. To some, technical solutions could be found, as in the case of spraying. Studies showed that effective spraying was hampered by the need, with existing methods, to mix the chemicals with large quantities of water which were not easily available on the farms. Here the introduction of ULV sprayers that do not use water offered a real solution (Norman 1974). In fact, the massive distribution of ULV sprayers was a principal ingredient in the World Bank sponsored "Agricultural Development Projects" which were launched in the main cotton growing areas in the mid-70s (World Bank 1973).

More generally, however, it was realised that peasant cotton was part of a complex enterprise, involving a number of other crops and activities with their own requirements in terms of resource allocation, labour time, and investment. Against the sole-cropping on the high-yielding experimental farm stood the mixed-cropping pattern on the peasant farm. Farmers did not adhere to recommend timing, for instance, because priority attention was given to food crops on which the household depended for its reproduction[6].

In 1968, representatives of the Empire Cotton Growing Corporation suggested according to Little that no significant increase could be expected in small-holder productivity. Nor was there much chance of the present area under cotton being significantly expanded. It motivated this consultant to recommend that "commercial production through plantations must be seriously explored" (Little 1969:64). It was necessary if the fast growing requirements of the Nigerian textile industry were to be met.

Other consultants were less sanguine on the issue of plantations and more optimistic about peasant production. T.S. Jones, investigating Nigerian cotton production on behalf of the Commonwealth Development Corporation argued that cotton plantations were unlikely to be viable economically. They had not proven successful either in Africa or in India (Jones 1969:45-6). As for the small holders, Jones maintained that "a doubling of the average 270 lbs seed cotton per acre which the farmers now obtain is well within their range of achievement" (p. 38).

The optimists about peasant production of the late 60s and early 70s were not vindicated by events. As stagnation shifted into deep decline, those looking for alternatives became more vociferous.

THE SHIFT TO FOOD PRODUCTION

By the mid-70s, the problem of competition from food crops had gone beyond the issue of its negative influence on the correct planting time for cotton. Cotton was increasingly marginalised in the overall cropping strategies of the farmers. The basic cause was obvious: The massive increase in the commercial demand for food as a result of the oil boom. The sharp rise in oil revenue and government spending, expecially after the 1973/74 price hike, was accompanied by a dramatic expansion in non-agricultural economic activity throughout the economy, in construction and manufacturing, in public and private services, transport and trading. There was a consequent outflow of labour from agriculture. Food prices rocketed and farmers responded by shifting whatever resources at their disposal in the direction of this new demand (Beckman 1987a).

For many farmers it led to an abandonment of cotton altogether. For others, it meant that scarce labour was concentrated on food crops, even if some land was retained under cotton. The shift to commercial food production has been recognised as a principal cause of the decline of cotton (NCB 1984)[7]

The shift can be illustrated graphically with the fate of the three World Bank projects (Funtua, Gombe, Gusau), designed originally with particular emphasis on cotton (World Bank 1973). As late as in 1979, the Cotton Board believed that project activities would lead to "a rapid increse in cotton production" (NCB 1979). On the contrary, as project management noticed where the wind was blowing, the focus on cotton was abandoned. The Bank projects threw in their resources behind the "maize revolution", thereby being able to "show results" (Beckman 1987a).

THE LABOUR PROBLEM

The shift to food production and in particular to maize in the principal cotton areas was closely linked to labour shortage and high labour costs. In the case of land, there was still scope for meeting the demand for food by putting new areas under cultivation. In the case of labour, farmers faced formidable competition from non-agricultural activities. Good prospects for wage employment also gave a big boost to educational investment. Farmers who had earlier been reluctant to send their children to school, were increasingly motivated to do so. The Cotton Board General Manager saw the inception of U.P.E., the Universal Primary Education programme, as critical for the fall of cotton (New Nigerian 11/11 1982). The spread of seclusionist tendencies within the Islamic communities which dominate the cotton growing areas also made female family labour lesse acceptable, in particular within the upper layers of the peasantry (Beckman, field work in Gombe 1980-83)

Farmers thus experienced a decline in access to family labour simultaneously as wage rates in the rural labour markets shot up in response to the expansion of employment in the economy as a whole. It also affected the terms on which farmers could draw on the labour of grown-up male children as paternalistic forms of labour relations were increasingly commercialised.

Cotton in particular was affected by these developments as it is a highly labour-intensive crop, requiring frequent weeding, and much labour for the picking. Women and young children have been a typical, extremely low-paid labour force for cotton harvesting. As this source of supply was undermined farmers' costs rose steeply. It has been recognised that it is the combined pressures of expanding food markets and declining labour supply that have hit cotton so severely (cf. Etuk a.o. 1985; IAR 1984:66-7).

Again, the comparison with maize, the new favourite crop, is striking. Here little weeding is necessary and harvesting is simple. It is also a crop that responded particularly well to the massive inflow of subsidised fertilisers that accompanied the World Bank schemes. As much of the new maize went to the booming poultry business, serving in particular the needs of higher income groups, the World Bank can be credited with the dubious achievement of having contributed to a drastic redirection of resources away from a key mass comsumer industry into such high-income consumption.

THE MARKETING BOARD

In the 1984 Cropping Scheme Meeting on the Fibre Research Programme (IAR 1984:67), it was suggested that "the major factor hindering the expansion of cotton was the lack of alternative market channels". This is a long-standing worry, ever since the state marketing system was devleoped after the second world war. At that time, all major export crops were placed under the control of marketing boards which were sole exporters and which appointed Licensed Buying Agents (LBAs) to purchase the crop. A fixed producer price was announced before the beginning of each buying season. The LBAs were rewarded by a fixed Buying Allowance per tonne bought. Produce Inspectors from the Ministry of Agriculture graded the quality of the crop (Onitiri and Olatunbosun 1972; Kriesel 1968).

The boards combined marketing, price stabilising, fiscal and developmental functions, often in a contradictory manner (Williams 1985). They were strategic in financing public sector expansion in the late colonial and early post colonial period. The LBAs constituted an important section of the emerging rural bourgeoisie. As agricultural exports declined and petroleum rose the importance of the boards also dropped. By the early 80s, almost all ran heavy deficits, paying producers above world market prices, that is, at official rates of exchange.

In 1977, regional marketing boards were reconstituted on single commodity lines (FGN 1977). The Nigerian Cotton Board was established with its headquarters in Funtua (NCB 1978). The history of the Board therefore coincides with the period of decline. It came under heavy attack from industrialists and farmers alike. Much of the critique was directed against the LBAs and their produce buyers. Farmers complained that they had to travel long distances and waste much time waiting in the cotton markets. They were often paid below the official prices and often by promisory notes which they found difficult to redeem. They were discouraged and preferred to grow crops which they could trade freely (Beckman, field work Gombe; CAP 1985a). Poor marketing arrangements were blamed for the fall in cotton production by the Sokoto State Governor (New Nigerian 17/5 1982).

The LBAs on their side complained that the Buying Allowance was insufficient. In the past the Board used to advance cash for payment to the farmers. Now the LBAs depended on their own finance and banks[8]. Some LBAs were not credit-worthy. Others obtained bank loans too late (NCB 1985b; cf. interview with NCB General Manager, New Nigerian 31/10 1981; also farmer's memorandum, New Nigerian 29/6 1985). Clearly, a fast shrinking crop and unchanged allowances were likely to reduce the cotton traders' interest in this crop, especially in a period of plentiful other opportunities.

Other complaints concerned the ineffective distribution of seeds. They were supposed to be provided free at widely dispersed cotton dumps. There were also problems with other inputs handled by the Board. The latter in turn complained of shortage of funds for performing its functions (NCB 1985a, b,c). This again was countered with suggestions that the cost of the marketing system was already too heavy (MANFU 1985; CAP 1985a).

In early 1986, the government announced its intention to dismantle the marketing boards. We shall return to this momentous move in a post-script.

THE PRODUCER PRICE

Ultimately, what mattered most was the price that farmers received. All the problems identified above depended on relative prices. Low yields, for instance, would not necessarily be a problem if prices were such as to make the effort worthwhile anyway. The same holds for the problems of competition from other crops and the high cost of labour. If prices were high enough to offset such extra costs, cotton would still be an attractive crop. According to IAR staff (Etuk, interview November 1985), little research had been done comparing production costs of different crops under peasant conditions. Such research is highly complex and full of pitfalls in the context of the varied inter-cropping strategies applied by the farmers.

Producer prices for marketing board crops - before the dissolution of the boards in 1986 - were determined by a federal Pricing Committee. The producer prices for cotton for the period 1969/70 to 1985 are given in Table 7. They are compared with production figures and an index for food counsumer prices. The latter is here taken as a measure, in the absence of better ones, of price developments in the economy as they may have affected food producers. There are no time series on actual prices received by the latter. There was a sharp increase in the cotton price in

1974/75 and 75/76, almost by 200 per cent. It can be seen in relation both to the sharp rise in wages at this point in time (the Udoji Award(and as a response to the

Table 7. *Cotton producer prices and cotton production*

	Production [a] 1000 tonnes	Producer price [b] naira/tonne (Grade I)	index (1975=100)	Composite consumer Price index: food[c] (1975=100)
1969/1970	91.4	112		
1971	39.4	112		
1972	38.4	124		
1973	48.4	134		
1974	30.9	113		
1974/1975	51.2	202	100	100
1976	58.4	308	152	122
1977	80,4	308	152	146
1978	37.1	330	163	172
1979	36.4	330	163	186
1979/1980	28.5	330	163	200
1981	26.4	400	198	250
1982	21.6	460	228	272
1983	19.9	510	252	336
1984	12.6	560	277	480
1984/1985	15.0	700	347	
1986	9.7	800	396	

Sources: [a] see table 6. [b] NCB Annual Reports. [c] FOS 1985.

poor crop of 1973/74, which was preceeded by a number of years of stagnant prices. The increase seems to have paid off in a marked up-turn in production. From 1976 onwards, however, cotton prices were allowed to stagnate. We see how they were fast overtaken by food prices. By 1980, the latter had increased by more than 60 per cent. The cotton price on the other hand had hardly moved. As late as in 1984, despite significant increases, it had only been allowed to rise some two and a half times over the base year of 1975, as compared with the five-fold increase in food prices. If we add farmers' difficulties in actually receiving the full producer price beause of trade malpractices the discrepancy grows.

It is not possible to estimate the relative importance of the price factor in explaining the decline of cotton, as other factors are not measured. It may be safe to conclude, though, that the problems of productivity, competition, labour and marketing recorded above were all reinforced by this unfavourable price development.

THE SHIFT TO IMPORTS AS A CAUSE OF DOMESTIC DECLINE

In seeking explanations for the decline in cotton production, we have so far looked at problems on the supply side: low yields, competition from food crops, high cost of labour, and problems of state marketing and state pricing. In understanding the shift from domestic to imported cotton, however, it is not good enough to stress difficulties at the agricultural end. The shift was propelled by forces outside agriculture. Most importantly, the expansion of the petroleum economy in the 1970s caused sharp domestic price inflation without corresponding adjustments in exchange rates. As a result it became cheaper to import than to buy locally. While the price of domestic cotton failed, as we have seen, to keep in step with over-all price developments, in the context of unadjusted exchange rates, the substantial increases in producer price actually granted to the farmers (see table 7, above) served to shift demand into imports, cotton or synthetics. To farmers, a better price was essential to compensate for high labour costs and other disadvantages that cotton suffered in comparison with food crops, expecially maize. Any additional price increase, however, only served to shift industrial demand away from domestic produce. Domestic cotton was therefore placed in an impossible situation were it was unable to compete for land, labour and other resources because the producer price was too low, while simultaneously unable to compete with imported raw materials because the price was too high.

Part of the problem was the open-door policy on imports practiced by the state which permitted this redirection of industrial demand from domestic to foreign sourcing during the heydays of the second (post-1979) oil boom. However, more effective protection of domestic cotton by higher tariffs is unlikely to have helped. The industry was equally undermined by heavy smuggling of final products which the state was incapable of preventing. Higher domestic producer prices and more protection would have further weakened the ability of the textile industry to defend itself against the smugglers.

As long as industry was allowed to import its raw materials freely it had little interest in pressurising the state to ensure the survival and expansion of domestic cotton production. The disintegration of the marketing board system was allowed to proceed without any outcry from the industrialists, those who ought to have had the greatest stake. As for the cotton farmers, while certainly bitter over the treatment meted out to them, they did not rise in protest. They did not need to because they had more profitable outlets for their efforts in other directions. They simply abandoned cotton.

We have argued elsewhere (Andrae and Beckman 1985) that imports of agricultural produce should not necessarily be seen as caused by a decline in domestic production but that the logic often operates the other way round: Imports cause domestic decline. On that occasion, the case was wheat, the massive importation of which we suggested greatly contributed to the stagnation of domestic food production in the 1970s and early 1980s. The case of cotton may be less obvious. But the same logic is at work. The problems confronting domestic cotton producers were real but the recourse to easy imports condoned by the state provided the textile industry with an escape route. The political pressures which should have brought about a reform of the marketing board system and more efficient services to the farmers were preempted. It is therefore misleading to put the blame for the decline of cotton on the farmers or on negligent Licensed Buying Agents, for that matter. The "negligence" of the latter was a function of a market where effective demand had been undermined by imports.

IV. Strategies for Raw Material Supply

RELUCTANT MANUFACTURERS FACE THE SQUEEZE

How did government and industry respond to the raw material crisis? The seriousness of the situation was only gradually realised. At the time of the National Workshop on Raw Materials in July 1983, the President of the Manufacturers' Association, MAN, was more concerned with import licences going to genuine manufacturers than with any direct involvement in production. Industries, he argued, could not take responsibility for producing their own raw materials (NISER 1983a:11). A committee on textiles set up at the workshop was even more explicit. The prospects for self-sufficiency in cotton production, as they saw it, were so bleak that importation was the best bet. They offered consolation in that "a lot of value" would be added locally to the cotton imported. A further advantage of liberalising cotton imports, according to the committee, was that self-sufficiency in cotton yarn (spun from imported cotton, though) could then be achieved. In their view, it was better to grow food and to import the cotton from the Sudan and Egypt (NISER 1983a:43, 46).

Two years later such a position was no longer politically feasible. Pressures had built up. As firms faced the paltry import licences allocated to them in 1984, they realised that the new government was as unable as the old one to offer any genuine relief. The government, on its side, sought to deflect the anger of industrialists and workers by pointing accusing fingers against industry itself. According to press reports, Idiagbon, the powerful Chief of Staff Supreme Headquarters, addressing the First Quarterly Luncheon of the Private Sector "urged industrial barons not to watch the factories close simply because there were no imported raw materials". They were directed to find local substitutes. Over the years, he said, firms had ignored government appeals for local sourcing. "The major firms, controlled from their overseas headquarters, have preferred to shop abroad because of the jobs and financial benefits it gives their home countries" (Guardian 31/5 1984).

In another statement to the business community, the Chief of Staff urged industrialists to resume full production and recall workers that had been laid off. To be able to do so the firms should go into plantation agriculture, which would not only provide food and raw materials but would "generate employment and check rural to urban migration" (New Nigerian 15/5 1985).

The manufacturers continued to press for more imports. In a MAN memorandum on the 1985 Federal Budget, the total inadequacy of current import allocations was deplored. At least 3 billion naira was needed in foreign exchange, according to the Association, in order to put industry back to a "reasonable level of capacity utilisation" (Financial Punch 13/3 1985). The Nigerian Labour Congress joined in the pressure for more imports. In an ambitious programme, "Towards National Recovery" (NLC 1985), the unions attacked import dependence. They also accused the firms of cheating with their licences, not using them for raw materials but for other business. They called for "complete government control and monopoly of foreign trade". A special agency should be set up for the importation of industrial raw materials. The NLC rejected the government policy of immediate local replacement as "not only simplistic but utopian", because it was no use appealing to the capitalists. They could only be moved by rewards and punishment.

In fact, the government was already brandishing both carrots and sticks. The Minister of Commerce and Industry announced, for instance, that only firms with a programme of sourcing a larger proportion of their raw materials locally or replacing imported components would be considered for approval (Business Times 4/3 1985). A Raw Materials Research and Development Council was to draw up time schedules for phasing out imports (New Nigerian 5/8 1985). In October 1985, the Ministry of Industries announced that an agreement had been reached with industry on such a time table. Firms were placed in two categories. They were given either five or ten years to comply depending on what type of raw materials they use. Agro-based industries were given five years (Guardian 21/10 1985). In the case of the textile industry, the five years applied to cotton, while synthetic fibres had to await completion of the relevant stage in the development of the petrochemical complex, officially scheduled for 1992. Industrialists kicked against the new time-tables which they considered unrealistic, especially in view of the inability of the government to release foreign exchange for the investment goods required for the industrial farms (Business Concord 17/1 1986). The deteriorating foreign exchange position following on the further fall in petroleum prices in 1985, however, reinforced the pressures on industry to find local solutions.

THE OPENING UP OF AGRICULTURE TO FOREIGN CAPITAL

An important element in the government's raw material strategy was the emphasis placed on foreign capital. The Minister of Agriculture presented a programme of incentives to encourage foreign investment, including a special tax relief, assistance in the acquisition of land, and special concessions with respect to the "indigenisation" laws. These were to be revised as to allow 80 per cent foreign ownership (Guardian 3/7 1984). Existing laws did not permit any foreign majority ownership in agriculture. Moreover, in no other sector was more than 60 per cent foreign ownership accepted. Agriculture was clearly singled out for special treatment.

The shift from 60 to 80 per cent did not materialise but, according to the Nigerian Enterprises Promotion Board, foreign firms were not deterred by the lower ratio. The Board reported a keen interest from foreign investors in the new deal. The US Agricultural Attaché informed the press that some 20 US companies were seriously looking into investment in Nigerian agriculture (New Nigerian 7/8 1985).

The new group of military men who took over government in August 1985 continued the same policy of invitation to foreign agri-business. Their President General Ibrahim Babangida, in a good-will message to the 25th Anniversary of the Nigerian-American Chamber of Commerce, had this to say in an official release headed "AMERICAN TECHNOLOGY NEEDED TO DEVELOP OUR RAW MATERIAL BASE":

> The American business community should be enlightened about the great investment opportunities and potentials in Nigeria. At present our major pre-occupation is to produce local raw materials for our industries. This is an area where the Americans can assist us substantially because they have the technology to develop our raw material base from the abundant resources within our country. They are also invited to invest in our agriculture and agro-allied industries... I see vast potential for cooperation between Nigerian businessmen and American entrepreneurs (Advertisement, Guardian 10/10 1985).

THE OPENING UP OF LAND TO AGRI-BUSINESS

Industry and its foreign partners were urged to go into plantation agriculture. Incentives were offered. The manufacturers, however, were not satisfied with the conditions. In particular they were concerned with difficulties in getting access to land, although the Minister of Agriculture had offered special assistance in this respect. State governments, he said, would be asked to make land specially available for large-scale farming (Guardian 3/7 1984).

The industrialists demanded the revision of the Land Use Decree of 1978. Although it had already allowed for much large-scale appropriation of land for state and private agricultural projects, it was ambiguous and subject to different interpretations. It could also, in fact, be used to obstruct such appropriation. The Kano branch Chairman of the Manufacturers' Association, when appealing for more incentives to go into agriculture, placed particular emphasis on the land problem. The Executive Director of the national Association accused the government of obstructing land acquisition. He quoted a case where the Ondo State Government had demanded excessive amounts to compensate present land owners (National Concord 29/4 1985).

In an "Interim Report on Raw Materials", the Manufacturers' Association suggested drastic remedies. Land should be leased for periods of not less than 50 years. A standard rent of three naira per ha should be charged and the rate of compensation for economic trees (the only compensation acknowledged in the Land Use Decree) should be fixed at 10 naira per ha (Business Concord 2/8 1985). The symbolic sums bordered on expropriation without compensation. Similarly, in response to government's attempt to tie them down to time schedules for local sourcing, the manufacturers made the revision of the Land Use Decree and the present land tenure system a central condition for their compliance with the deadlines (Business Concord 25/10 1985).

Industry joined the rising agrarian bourgeoisie in an intense psychological warfare to force open the remaining barriers to the full commercial appropriation of land, while simultaneously using special connections to local rulers to get privileged access on non-commercial terms. Most military governors pledged assistance. Governor Mohammed Umar of Kwara, for instance, offered to donate land free of charge (National Concord 14/11 1985). Governor Chris Garuba of Bauchi directed the State Land Allocation Committee to accord priority to requests for industrial land and large-scale farms (New Nigerian 18/1, 29/1 1986). The Ondo Governor, Michael Akhigbe, promised that land acquisition would be no problem (Guardian 12/11 1985).

The Governor of Kaduna, Abubakar Umar, however, warned that multinationals were using the sourcing of raw materials as a pretext for land speculation. His government, he said, would not tolerate the massive displacement of peasant farmers. He insisted that companies should obtain their raw materials from peasant farmers and not by dislodging them from their source of livelihood (Sunday New Nigerian 23/2 1986). Governor Yohanna Madaki of Gongola was similarly quoted as saying that his government would no longer tolerate a situation where 99 per cent of the people were made mere tenants on their own land by a few privileged individuals (Guardian 7/4 1986). He claimed, according to this press report, that no state in the federation showed such a degree of abuse of peasant land rights as Gongola, necessitating the dissolution of the Land Use Advisory Committee. Interestingly, as we shall see below, the first two major industrial cotton plantations were both established in this state. The land issue was no doubt politically explosive.

AN IMPORT SUBSTITUTION PLAN FOR COTTON

How did the new raw material policies affect the textile industry? As part of the "agreements" with the government in 1985 (cf. above), the industry committed itself to eliminate the importation of cotton over a five years period. Imports of cotton yarns were immediately banned. The NTMA submitted a programme for local sourcing including a year-by-year phasing-out plan (NTMA 1985b). As we saw in chapter II, (Table 5), the industries' requirements for cotton materials were 75 - 100,000 tonnes per annum at a total material use of 125,000 tonnes, depending on the level of substitution for man-made materials. We also saw that less than 20,000 tonnes have been domestically supplied since 1982.

The import substitution programme was based on an assumption of an annual increase of 100,000 bales or 18,000 tonnes, that is, equivalent to the whole domestic supply at the point of departure. It thus implied a virtual five-fold increase over the five years period! The cotton industry was expected to build itself up in this short period to a level 20 - 30 per cent higher than its unique peak in the 1969/70 and 76/77 seasons, or to a level twice the average achieved during the 60s and 70s.

The NTMA must have realised the blatant unrealism of the assumption behind such formidable recovery *cum* expansion. But the assumptions were not their own. The Association could take cover behind projections offered by the Cotton Board. The Ministry had apparently directed the NTMA to use those projections as a basis for the plan submitted to the government. This plan was therefore not a "substitution" plan as such but merely a theoretical exercise suggesting how imports could be phased out, *under the assumption* that such domestic supply would be forthcoming. How it was expected to do so was not spelt out; nor were the assumptions underlying the Cotton Board's projections.

THE ROLE OF THE COTTON BOARD IN THE REHABILITATION OF THE COTTON INDUSTRY

How was this "great leap" to be achieved? Let us first look at the position of the Cotton Board, the institution most directly responsible for the welfare of the cotton industry. Spokesmen of the Board were anxious to emphasise that the Board was just a marketing organisation and therefore not responsible for production as such (interviews, NCB, Funtua, October 1985). Yet the interpretation of its mandate and the ambition to play a role in production varied from time to time. As the cotton crisis deepened, the Board considered direct large-scale investments in plantations of its own. Plans were announced in 1982 for altogether some 40,000 ha of plantations, to be distributed equally over the ten northern states, drawing on foreign technical partners (NCB 1982b). Negotiations for land were undertaken with the various state governments. Three years later, these plans had been shelved. Nor did another project, for which a feasibility study was prepared by foreign consultants (Man-Tech), result in any concrete action. This inactivity reflected the tendency of the post-Shagari regimes, to oppose direct state involvement in agricultural production, including the attempt to privatise public enterprises in this field. The River Basin Authorities, for instance, were instructed to withdraw from direct production.

The Cotton Board continued to maintain that it was not possible to depend solely on small farmers in view of the decline of production. In a memorandum on the "Causes for the Decline of Cotton and Some Measures to Remedy the Cotton Situation", a spokesman of the Board emphasised the availability of land for such large-scale ventures but pointed to the problem of the large capital required to achieve the level of mechanisation necessary to overcome the labour bottle-neck (NCB 1984b). The memo suggested the establishment of a system of nucleus farms, including a central "company" farm, and farm service centres providing inputs and services to "outgrowers". Foreign technical partners and federal finance were part of the assumptions. The abolition of the marketing boards in 1986 put an end to such plans.

The principal area of Cotton Board involvement was in marketing. An increasing emphasis was placed on big farmers. A special scheme was introduced in 1982 for loans to large-scale farmers, defined as those with 30 ha and above (NCB 1982). The failure to recover outstanding loans in combination with deteriorating Board finances led to a suspension of the loans scheme already in 1983 (NCB 1983). In the 1985/86 season a new scheme of incentives for large-scale farmers was introduced. They were to be given an extra "allowance" of 40 naira per tonne on top of the producer price. Their cotton was also to be graded and purchased by the Board directly on their farms (Business Concord 4/10 1985). The incentives were justified in terms of economising market operations as it was cheaper to handle a few big parcels than many small ones. The bias of the scheme was still in line with the general assumption that the "typical" small holders of the past could no longer be relied upon.

The Board's plans for the "rehabilitation" of cotton focused on the need to build more permanent markets and the free distribution of insecticides and herbicides to "deserving farmers". Its main effort was to secure finance for a more ambitious programme in this direction. Hope was pinned on the textile industry coming to the rescue, especially as many of the textile mills continued to make profits (NCB 1985b). While pleading for more funds for its own rehabilitation programme, including direct subventions from the textile industry, the Board increasingly acknowledged the new role of the manufacturers in the development of the cotton industry:

> Textile mills may set up large mechanised production units to supplement the production of the traditional farmers and/or involve themselves very actively in agricultural extension and supply and support services (NCB 1985b).

In this respect, the Board fell in line with the prevailing political mood in ruling circles of Nigeria. Partly, it was an attempt to shed direct responsibility: Let industry take care of its own raw material problem! But there was another dimension to it: industry as the saviour coming to resurrect moribund agriculture.

The Board was supported in this respect by the cotton ginners (CAP 1985a). They wanted the Board's functions to be reduced to those of a watchdog and instead involve the textile industry as well as the ginneries in marketing, nucleus farms and group farming schemes; drawing on pools of mechanised services and other inputs provided by the textile industries, either collectively or on an individual company basis. The Chairman of the "Committee on Cotton Production in the Ten Northern States", also came out in favour of such schemes for linking up farmers and industry. (New Nigerian 8/5 1985.)

TEXTILE MILLERS STRUGGLE FOR CONTINUED IMPORTS

The first choice of the textile industry when it came to securing its raw material supply continued to be importation. It was cheaper, due to high Nigerian production costs both in cotton agriculture and in spinning and weaving. This was further exacerbated by the over-valued naira. In the current shortage situation the discrepancy was even greater: Millers spoke of a difference of over 100 per cent between the cost of imported and local yarns (interviews July to October 1985). Since 1984 cotton prices to millers as set by the Cotton Board were separated from relation to the world market price in order to reflect real, that is, much higher costs (interview Udokwu 16/10 1985).

Fierce struggle for the scarce remaining import opportunities was thus the immediate reaction to the import squeeze. The press reports referred to in chapter II with warnings of great problems to come, were all related to the life-and-death necessity for adequate licences to import. The struggle was both between competing firms and between textiles and other sectors. The particular pressure put by government on the agro-based industries to achieve local sourcing was met by pleas to consider the labour-intensity of the textile industry as an argument for more generous treatment. The sector Association, NTMA, watched over sectoral interests and regulated internal conflicts as coordinator and party to negotiations with the Cotton Board on licence allocations. These were based on the stated requirements submitted by individual firms. The gross overstatements in the aggregated figures reflected the competitive situation and was cause for some concern by the Association which had to mediate a "fair solution" for all its members (interview Young-Itiye 13/9 1985). The statements were treated with sanguinity by the allocating Cotton Board. In the main allocation in March 1985, the principle was applied that 50 per cent of the shortfall between stated requirements and local supplies were recommended for imports (interview Udokwu 16/10 1985).

There was an obvious source of conflict here. In the January 1986 Federal Budget, the government announced a new import policy where licences were to be allocated in bulk to each industrial sector (Guardian 14/3 1986; Business Concord 11/4 1986). Sector organisations would themselves be responsible for allocations among their members. The proposals led to a rush of applications for membership (Business Concord 25/3 1986). In the case of the textile industry, the NTMA set up a Technical Committee for the purpose of gathering data that could provide a basis for the allocations (Guardian 7/ 1986). The scene was set for bitter infighting among members, using regional and other sectional platforms to enhance their bargaining positions. In the case of the allocation in April 1986 mutual recriminations were fierce. NTM, a leading Lagos firm, for instance, claimed that it had been grossly disfavoured and that the government had approved more than 22 million naira worth of licences to 56 "ghost" textile firms (Guardian 8/5 1986). The new policy of sectoral self-management, however, was soon to be dropped as the government decided to abandon licencing altogether in favour of the auctioning of foreign exchange under SFEM (see p. 55 below).

BACKWARD INTEGRATION: A LONG TERM SOLUTION

The NTMA programme for phasing out imports by 1991 was premised on the continued ban on the importation of cotton yarn. The substitution plan for man-made materials in the same programme assumes the continuous substitution of the imported man-made yarns by locally produced fibres and yarns. This was in anticipation of the petrochemically linked production of the polymer-"chips" on which these fibres are based that was programmed for 1992. The assumption was therefore for great increases in both yarn and synthetic fibre production. Provision of adequate spinning capacity is indeed a major challenge to the textile industry when it comes to achieving the import substitution goal as a whole. The NTMA programme assumed an expansion from 106,000 to 128,000 tonnes from 1986 to 1991.

We would rather depart from from the level of spinning done in the 77-81 period, as indicated by fibre supplies available, and assuming this to reflect the level of currently available capacity. As shown in Table 4, this level is at about 50,000 tonnes. Assuming the same structural relations as above, 60 per cent cotton - 40 per cent man-made, we too, arrive at an additional requirement of 25,000 tonnes spinning capacity for cotton yarns if the 125,000 tonnes production level is to be achieved. But then there is also the substitution of man-made imports to take into account. If we continue to assume a 50 per cent substitution level of the man-made component at this production level, another 25,000 tonnes spinning capacity is required. Expansion from 50,000 to 100,000 tonnes, i.e. a doubling of capacity, will take quite a concerted effort. What has happened?

Numerous announcements in annual company reports and in the press were made in response to the policy for expanded spinning. Our interviews indicate further plans in this direction. Afprint Nigeria Ltd and Five Star Industries Ltd, both Lagos based firms with Indian participation, announced new spinning installments in early 1986. In the case of Afprint they were to take place in a new subsidiary company, Aflon Nigeria Ltd, to use modern rotor spinning technology to produce 3,000 tonnes of yarn per annum. The external content of the 22 million naira investment was to be borne by Nigerian Industrial Development Bank,

NIDB, which already has equity in 15 textile firms and 20 per cent of that in Afprint. A credit line offered by the World Bank was to be used. (Business Concord 22/4 1986 and interview Maira 10/9 1985). Enpee and President Clothing Company, also Lagos-based, Indian-controlled firms, in a joint project for a new spinning company, United Spinners Nigeria Ltd., present another case in point (Guardian 22/1 1985, and interviews Pahuja 9/9 1985 and Ibie 20/11 1985).

A particular challenge is the building of capacity for production of so-called filament yarns (non-spun synthetic yarns). Only a small capacity of 5,000 tonnes existed in the country, in one company, Nichemtex in Lagos. Plans for a plant intended to be collectively financed by several industries in cooperation with some state funds and spearheaded by the Nigerian Industrial Development Bank, however, seem to have met with little response from industrialists They were said to prefer to do their own individual investments (interview Ajayi 12/9 1985). Quite apart from the spinning challenge there is again the task of establishing the capacity to make the man-made fibre from the polymer chips. Capacity has existed since 1975 to handle this processing stage, but only for some 5,000 tonnes of fibre. It has also been based at Nichemtex. A new venture, the Polyfibre Company Ltd. established in Kaduna in late 1985, was however a sign that also this line was moving. They are to produce carpets from materials from the NNPC refinery at Kaduna. Interest for collective provision via NIDB was greater in the case of this process, according to Mrs Ajayi.

In terms of import substitution these have all been long-term projects. They depend on access to foreign exchange for machinery investments. As a strategy of backward integration it could be featured by the industries as responsible compliance with national efforts toward solving the raw material problem also in these respects. The ban on yarn imports in favour of cotton fibre reflects the strength of the large integrated firms that dominate spinning and also control the NTMA (e.g., KTL, UNTL and NTM). It will have been in the interest of the non-spinners to rapidly regain control over supplies by investment in new spinning capacity, in the ensuing oligopolistic market situation of great shortage and very high yarn prices. Theirs was the most precarious situation. Up until the end of 1985 it was however still possible for them to expand capacity in non-cotton processing. The NTMA chairman was bitterly complaining of the continued allocation of import licences for materials and machinery for such ventures (interview Ilo 20/11 1985).

THE TEXTILE MILLER AND THE REHABILITATION OF COTTON PRODUCTION

Reliance on the marketing board system had meant that the Nigerian textile industry had little direct responsibility for (or control over) its own raw material supply. This changed as a result of the shortage situation. One of the first commitments made by the textile industry towards encouraging cotton production was to contribute 20,000 naira collectively towards a fund in support of the general work of the Cotton Board. This was made through the NTMA in 1984. The distrust of the efficiency of the Cotton Board was however widespread. Criticisms of the LBA buying system and the pricing policy abounded. Calls were made for the dissolution of the Board for failing to cope with declining production (New Nigerian 15/4 1985). Statements such as "They have outlived their usefulness",

"They have been given their chance" were recorded in several of our management interviews.

The pressure was building up inside the industry to take matters in its own hands. One model was for industrial "end-users" to take over from the Cotton Board the responsibility for marketing and support functions in the contact with the cotton growers. It was propounded with some force by another group of "end-users", the Cotton and Agricultural Processors, the company that did most of the ginning for the Cotton Board. This view was contained in a memorandum first prepared by its parent company, the British Cotton Growers' Association, in early 1984, which was later submitted in an up-dated form by CAP to the Shettima Mustapha Committee (CAP 1985a).

Rejection of the Cotton Board's way of handling its task did not necessarily mean rejection of the reliance on small-scale peasant farmers as producers. The CAP in its memorandum went out of its way to emphasise the necessity to make use of existing production structures. The main point was that marketing should be controlled by industry for greater efficiency. Explicit inspiration for the model was drawn from the example of the Nigerian Tobacco Company, NTC, in its successful procurement of tobacco, achieved by far-reaching control over both input distribution and purchasing by the industry itself. CAP pictured similar forms of control by a group of textile companies.

The NTC model was taken up and elaborated by the United Nigerian Textiles Ltd., UNTL, the largest company of the conglomerate of spinners, weavers and printers centred in Kaduna, in a concrete proposal backed by a consultancy report presented in March 1985 (MANFU 1985). The main suggestion was for the formation of a company, where all the textile mills would be required to purchase shares. Ginners, (seed crushing) oil mills and also the Cotton Board would be asked to join. The company was proposed to replace the Cotton Board in the three main areas in the North where 90 per cent of all cotton is grown (Bauchi, Sokoto and Kaduna States), leaving only some overall tasks including seed multiplication to the Board in these areas. The new company would operate by a hierarchically organised input distribution and marketing organisation with spraying, tractor hiring, and demonstration services to offer. It would also have research functions and demonstration farms. The base of the hierarchical organisation would be groups of farmers led by "contact farmers" of which 4-5 would connect with one of some 400 extension agents from the organisation. The equity to be contributed by the industries would amount to some 5 million naira, in addition to contributions to finance purchasing.

The UNTL proposals followed closely the NTC model. Some main differences in the underlying conditions were mentioned but not discussed. One relates to the scale of the operations. Whereas NTC has to deal with 50,000 farmers to produce 14,000 tonnes of tobacco, the cotton company would have several hundred thousand farmers expected to produce some 85,000 tonnes of produce. Secondly, the NTC has built up its organisation from scratch over more than a generation and has had ample time to adjust and learn by experience. The UNTL plan was for five years in the first phase. Finally there was the obvious difference of one single company being in control in the case of NTC, against the proposed conglomerate of almost one hundred textile millers.

The UNTL proposal was not adopted by the NTMA. When in August 1985 the Association met to consider its further plans it rather decided on a double strategy for the next ten years which implied continued short-term support for existing marketing structures in combination with a definite element of long-term commitment in the direction of large-scale mechanised plantation farming (interview Abubakar 4/10 1985). The financial commitment was considerably stepped up. By raising contributions from member companies they proposed to collect 3 million naira annually. To facilitate contributions to the joint fund, the chairman of the Association called on the government to refuse import licences to textile firms that failed to cooperate (Business Concord 4/3 1986). The NTMA also threatened to expose those unpatriotic companies which "have gained much from Nigeria in the past and are now refusing to help her economic recovery" (New Nigerian 14/5 1986). The commitment to large-scale mechanised farming was expressed in the proposed support for a new research programme to be carried out at the Institute for Agricultural Research at Zaria. A memorandum from the IAR to the NTMA outlined the areas of research required in order to improve the information concerning conditions for the introduction of large-scale mechanised farming of cotton in Nigeria (IAR 1985). It specified projects for which the Institute would welcome NTMA support. While it emphasised the continuing importance of the peasant system, it assumed that to step up production as much as was now required, an increase in the total land cultivated was necessary. This again, was assumed to require mechanization.

The IAR emphasised the bottle-neck faced in harvesting. Existing cotton picking machines were highly sophisticated and geared to very large-scale production. Apart from the very long-term task of finding simpler small and medium scale solutions, there was the need to develop cotton varieties of the uniformity required to suit mechanical picking and to master the required adjustments in other aspects of farming such as weed control.

The IAR proposals maintained that small farmers should continue to be encouraged by "all possible means". They justified the research into large scale production, by the lack of any knowledge in the area. They were also careful to point out that "mechanised large-scale production in Nigeria is not likely to be an easy venture and should therefore be introduced rather gradually and systematically, until such time when all the necessary resources (both human and material) for this type of system become readily available".

There were no necessary contradictions between the UNTL/NTC strategy and the interest in large-scale plantation farming. The possible combinations were many. The UNTL proposal discussed with interest block farming and nucleus estate models, where plantation type of production is integrated with small-scale peasant farming. The main force behind the commitment to research into problems of mechanisation, however, was likely to come from those companies which had already demonstrated their interest in the large-scale plantation solution by actually acquiring land.

Three cases of direct agricultural investments by textile mills were mentioned in the introduction: President Clothing, Afprint and Nigerian Textile Mills (NTM), all Lagos firms. A few others have also been announced: Texlon, Lagos, together with its sister company Terytex in Kano, are reported to have acquired 700 ha in Bauchi for large mechanised farming "to serve as a demonstration farm to the neighbouring farms to encourage them into cotton production" (Business Concord 9/8 1985). Chellco Industries, a blanket factory in Kaduna, Aswami Textile Industry, Lagos and United Textiles Industries are all companies that are associated with the Chellarams and Sons (Nig.) group of companies. The latter has established a subsidiary, Chelagric (Nig.) Ltd., with 200 ha farmland, and plans to acquire a further 1,000 ha in Niger State (New Nigerian 10/6 1985). The Bhojsons Group of Companies, another Indian controlled chain that includes Globe Spinning Mills has similarly created a new subsidiary, Globe Agroventures, "pioneer project for plantation of cotton and other farm produce" (Advertisement on India's Republic Day, New Nigerian 26/1 1986).

An interview at Enpee Industries in Lagos (Pahuja 9/9 1985) revealed far-gone plans for an agricultural venture in Gongola State, to grow cotton among other crops (e.g. tomatoes). A few other cases were mentioned by the textiles cotton community to be in the pipe-line. Cotton Board officials confirmed that many firms had approached them for assistance to identify and acquire good available cotton land (interviews Wadu 9/10 1985 and Kankia 16/10 1985).

Many of these announcements represented plans which may not be easily realised. The problems faced by the new industrial farmers were many. As for other industries the access to land caused frustration. Representatives of the Textile Institute and Afprint were quoted in press reports from a symposium for textile manufacturers in April 1985 to have complained about the obstacles the land issue was causing to the efforts towards backward integration, especially the complications caused by the Land Use Decree (Guardian 12/4 1985). A major barrier referred to by millers was the size of the capital that had to be committed at the very time when the raw material problem, combined with earlier crisis elements (see above), had reduced the funds available: "The investment in a viable cotton field is tremendous, and is really something that most of us are not financially strong enough to do right now", said Mr. Ilo of NTM and the NTMA in an interview published in September 1984 (Ilo 1984). He added that in principle it was an interesting undertaking. One year later his company was the one that had committed the largest investment in this direction.

The NTM planned for a 20,000 ha plantation at Gassol in Gongola State. They joined two other companies to form Associated Cotton Growers Ltd. The senior partner seems to be SCOA, a major French controlled conglomerate, a colonial merchant firm that has successfully diversified into a wide range of trading and manufacturing, including vehicle assembly and textiles. The SCOA Chairman, Ahmed Joda, a former top bureaucrat ("Super Perm Sec") and experienced businessman with major private agricultural interests in Gongola, is likely to have been instrumental in securing local support. Newspaper reports suggest that another partner was Temperance Farms Ltd., a company owned by the former Head of State, General Obasanjo (Business Times 26/1 1987). The NTM claimed that it was committing 23 million naira to the venture, half of which would be spent in the first five years (Guardian 24/10 1985). A small pilot farm had already been planted in 1985. Some 2,500 ha would be developed every year. Actual planting in 1986, however, was reported to be 1,000 ha (Business Times 27/10 1986).

The NTM project in Gassol was lagging behind the other major new industrial cotton plantation, also in Gongola, the one operated by Afcott, a subsidiary of Afprint, at Ngurore near Yola. We visited this project in July-August 1986. Here is a report.

AFCOTT: THE PIONEERS

As we arrived in Ngurore, the first cotton crop was being planted. An ex-Rhodesian pilot, now based in Zimbabwe, was busy spraying the fields with a small airplane, bringing home to us, instantly and graphically, both the transnational nature of the venture and the sophisticated level of the technology involved. The cotton operations at Ngurore are handled by MASDAR, a British based management and consultancy company with wide-ranging interests in Nigeria. They act on behalf of Afcott, a subsidiary of Afprint, one of Nigeria's biggest textile firms. Afcott rents land from the Upper Benue River Basin Development Authority, a federal development agency which has been engaged in agricultural projects in the area since the late 70s (Amale 1986). Afcott's cotton plantation is located on the Mayo-Ine Project, the single largest run by the Authority. So far Afcott has taken up 1,700 ha of which 1,300 is cultivated directly by the company. The remainder is allocated to "outgrowers", that is, local villagers, on the basis of one ha per person, according to the Project Manager Mr. Knaggs, a former landed estate agent with a background in Kenya, Rhodesia and Florida.

The mother company, Afprint, is Lagos based since incorporation in 1964. It began producing prints from imported grey baft. It later expanded into weaving and spinning (Nigerian Enterprise 1984). The principal owners are of Indian origin, Chanrai, a dominating family in the textile industry in Lagos. Their share was reduced to 55 per cent after the introduction of the 1977 Indigenisation Decree. By 1985 it was 45 per cent (interview Maira 10/9 1985). As with many other textile firms, Afprint went through a difficult time in the late 70s and early 80s. Major restructuring, however, led to improvement in profits in 1983-85, despite the deepening national economic crisis (Afprint 1981-1985). The company was therefore in a position to finance new investments as the raw material problem became acute. We have cited above its establishment of a subsidiary company for expansion of its spinning capacity, Aflon Nigeria Ltd. in 1986.

Plans to go into agriculture were announced at the Annual Meeting in 1985 (Financial Punch 25/9 1985). Press releases in early 1986 spoke of a first stage of 2,000 ha, capable of satisfying 20 per cent of Afprint's cotton requirements. The farm was to be expanded to 10,000 ha within a five years period at a cost of 30 million naira. 3,000 workers would be employed during peak periods. Subsidised inputs were also to be provided for small-holders (Business Concord 24/1 1986).

Afcott was advertised as a "giant leap for the textile industry" and as "one of the largest single cotton plantations in black Africa" (e.g. New Nigerian 20/2 1986).

Linking up with Upper Benue River Basin Authority provided a fast and handy solution to the land problem. Afcott, as other new industrial farmers, felt ill at ease when dealing with peasant land owners. A company executive complained that "you have to drop everything you have to the man who will give you land" (Guardian 12/4 21985). He felt recent land legislation made things more difficult by forcing prospective buyers to negotiate both with individual owners and with the government. In the case of Upper Benue, however, the Authority had already "solved" the problem as the Mayo-Ine lands had been taken over by the state for "river basin development". According to the Upper Benue Project Supervisor, some 38,000 ha of cultivable land had been acquired (interview Santuraki 1/7 1986). Some 2,000 ha had been prepared for cultivation annually from 1982, but the development had been brought to a halt due to the financial crisis plus a federal directive in 1984 ordering the River Basin Authorities to withdraw from direct production, in line with the new "privatisation" policies. Land clearing equipment had broken down because of lack of spares. Upper Benue was therefore anxious to find private companies willing to move in, both to take up land already cleared by the Authority, and to develop new land. Afcott was consequently spared both the problem of land clearing and that of direct confrontation with the peasant land owners. The company, however, also looks for land of its own, with greater security of tenure (interview Knaggs 30/6 1986).

No information is available on the size of the peasant population in the area before the arrival of the project. The cultivated area at any one time may have been small. But the disruption of the peasant economy is still likely to have been great because of the system of land rotation and fallow, in combination with a large cattle population. Initial confrontation with the peasants was lessened as part of the "project land" was given out on short-term lease to the villagers. But as big companies were moving in, either in joint ventures with Upper Benue or as tenants, the scope for land distribution declined. Much of the land taken up by Afcott had been farmed by individual small-holders in previous season, according to the Project Manager.

Apart from its own plantation, Afcott operates an outgrowers' scheme. Land leased by Afcott is again leased to small-holders who are supposed to receive cotton seeds and other inputs from the company. The outgrowers sign an agreement to grow cotton, follow project advise, and to deliver the produce to the company. If they don't, the farm can be taken over (Afcott 1986). The outgrower, contract-farming component however seems to be marginal to Afcott's overall project design, at least for the time being. Only 430 of the 1700 ha planned for cultivation in 1986 were given to outgrowers. It is not a "nucleus farm" model where a central plantation forms a basis for supporting cultivation by small-holders in a vast hinterland. In the present lay-out, the outgrower areas are restricted primarily to narrow strips around the villages which have been encircled by company land. In theory, the company is supposed to provide mechanised services but no land preparation had been undertaken on outgrowers' land when we paid our visit. Field operations on company lands, on the other hand, were far advanced. The combination of extreme insecurity of tenure and little company support makes these outgrowers unlikely candidates for the dynamic and productive role attributed to "contract farmers" in recent literature (Buch-Hansen and Secher Marcussen 1982).

Why this emphasis on plantation agriculture? Both the Farm Manager (Hancock) and the Project Manager (Knaggs) thought that small-holders would in fact produce cotton more cheaply. The latter, however, argued that the profitability of the plantation was ensured because of its integration with industry. Farm operations as such may not be profitable but in a situation of extreme shortage of cotton any controlled access would pay off as it allowed for the activation of industrial capacity that would otherwise be idle. Physical control over supplies, not the relative productivity of different farming systems, was what mattered. The Project Manager envisaged a future point in time when commercial supplies of cotton would be adequate with no need for direct involvement by industry in production.

The primacy of controlling supplies was further emphasised when we discussed Afcott's plans to establish its own ginnery despite the fact that there is much idle or underutilised ginning capacity in the region. Apart from ginning Afcott's own cotton, the new ginnery would serve as a vehicle for controlling access to small-holder cotton in a wider area in a situation of cut-throat competition for supplies.

Why would plantations not be profitable in their own right? Apart from the general problems of estate management in the Nigerian context, a particularly heavy burden is imposed by overhead costs in a situation where management is imported. At the time of our visit, there were five European management staff, including three farm managers and an engineer. There is also heavy dependence on a type of technology that is imported, including airplanes and chemicals. The mechanisation of harvesting was also envisaged and two cotton harvesters had been ordered as a first step in this direction. These are highly sophisticated machines which depend for the realisation of their productivity potentials on strictly controlled production methods and special cotton varieties. From the perspective of foreign management, however, there was the primary need to reduce dependence on local labour. Mechanised harvesting would lessen the problem of supervising and disciplining vast numbers of casual labourers, especially in an area where Islamic influences restrict access to more easily controlled female labour.

The production system developed by Afcott in Ngurore can only be understood in terms of privileged access to scarce foreign exchange. It has been the policy of the Nigerian state to facilitate such access for companies which have gone into local sourcing of raw materials. However, even without such policy, companies like Afprint and Afcott are well placed to ensure access, due to their position in Nigeria's transnational political economy. The prospects of Afprint's venture into cotton production will thus continue to depend on the ability to sustain this model in a context of changing foreign exchange regimes. When SFEM - the Second-tier Foreign Exchange Market - was introduced in September 1986, the mode of access was drastically altered. Prospects also depend on the ability to draw on continued state backing for special access to land, at the expense of local peasant communities. Such politically determined advantages may not in themselves generate a more productive farming system, that is, a system that is more "cost effective" than independent small-holder production. The superiority of the system, however, as seen by the manufacturers, lies in the way in which it ensures direct control over cotton supplies already at the stage of primary production. It allows Afprint to “internalise” trading profits which would otherwise have accrued to “middlemen”, in a highly protected market where there is an extreme imbalance between demand and supply.

V. The Scrapping of Marketing Boards and the New Foreign Exchange Regime of 1986: A Post-script

MARKETING WITHOUT MARKETING BOARDS

The relationship between industry and agriculture took a radically new turn with the scrapping of the commodity boards in 1986. State participation in agricultural marketing had been a major target of attack by international reform strategists, especially since the Berg Report (World Bank 1981). The abolition of the boards was high on the agenda of the World Bank when negotiating its financial support for Nigeria's "structural adjustment programme". As the financial position of the Nigerian state further deteriorated during 1985 as a result of the decline in petroleum prices, the boards were natural candidates for cuts. In the case of cotton, we noted that the textile industry (as well as cotton farmers and licensed buying agents) had complained bitterly about the performance of the board. It did not necessarily mean that the industry as a whole was in favour of abolition. Views differed strongly among the firms. But demands for reform were frequent from all quarters.

The decision to scrap the boards was announced in April 1986. Crop buying was to be discontinued from 1 June and the boards were given until the end of the year to wind up. The Federal Minister of Agriculture in briefing the press spoke of the massive losses incurred by the boards, their inefficiency and their negative impact on crop developments (Guardian 24/4 1986). From now on state ministries of agriculture would take on the principal responsibility for crop development. There would still be federally recommended minimum prices but the state would only enter the market in the last resort.

The decision set the stage for a new chaotic phase in the relationship between the textile industry and the cotton producers, the full consequences of which have yet to unfold themselves. There was at first no uniform pattern as each of the governments in the major cotton growing states settled for different strategies in dealing with the new situation. The NTMA, through its Vice-chairman Malam Abubakar, assured the farmers that the Association would buy all the cotton that they could produce (New Nigerian 14/5 1986). The Association saw itself as the principal successor of the Board. It planned for the establishment of a body with a responsibility to buy the entire cotton crop while liaising with the various state governments which had been entrusted with responsibilities for crop development (New Nigerian 10/7 1986).

Negotiations were initiated for the taking over of the assets of the Cotton Board. In October, as the new cotton harvest was approaching, the NTMA launched the Nigerian Cotton Company Ltd. to be the only buyer of cotton on behalf of the Association. The company would appoint its own buying agents. The cotton was supposed to be shared among the members according to an "acceptable formula" (New Nigerian 11/10 and 13/10 1986). The Association offered to buy cotton at 1200 naira per tonne, a two hundred naira increase on the government suggested minimum which in turn was 150 naira up on the price of the 1985/86 season. The new company, according to its sponsors, would provide an effective alternative to the Board in terms of "providing all the services formerly rendered by the latter". It would be fully funded by the textile manufacturers to ensure timely payments to farmers. It would provide the machinery for pursuing the rehabilitation of the cotton industry (Abubakar 27/1 1987).

Cotton buying began in earnest in December. It was soon claimed that a bumper harvest was in the making. Reports from all major areas suggested at least the doubling of the 1985/86 crop. But there was no jubilation. On the contrary, there were widespread reports of aggrieved farmers pledging never to grow cotton again because of the difficulties they faced marketing their crop. The market was in disarray. Farmers claimed that there were no buyers and that they were forced to sell at much below the government minimum (New Nigerian 12/1, 16/1, 30/1, 9/2 1987; Today 8/2 1987; Guardian 18/2 1987; National Concord 27/2 1987). A visit to Gombe, a key cotton producing area, in February 1987, and discussion with farmers and traders confirmed these difficulties and disappointments on both sides. The crop had at least doubled according to ginnery sources, but much was left unsold or unpaid for. What had gone wrong with the bold pledges of the textile manufacturers?

Malam Abubakar, the NTMA Vice-Chairman and the Chairman of its Raw Materials Committee gave his own version in a bitter memo. "Brief on the State of the Cotton Industry in the Aftermath of the Dissolution of the Nigerian Cotton Board" (2/1 1987). The disarray of the cotton industry, he argued, was due to the failure to create a satisfactory alternative to the Board. This again was because of the suddenness of the decision to dissolve it, the lack of consultation with the affected parties, and the failure to provide any transitional arrangement. In the past, the Board had advanced funds to its buyers. The new buyers soon ran out of funding. Operational materials (bags, twines, tarpulines, etc) were locked up with the Board whose assets remained to be disposed of. Most seriously, no arrangement was made for ensuring that seeds for next years' planting were taken care of for treatment and distribution to farmers. Buyers were irresponsibly selling seeds to the vegetable oil mills for crushing, according to the NTMA chief.

What happened to the Nigerian Cotton Company (NCC) which was to replace the Cotton Board? Most members failed to cooperate with this NTMA scheme. Major firms settled for separate buying arrangements. Others chose to import. Sani Ismaila, a manager in the UNTL group with one-fifth of the country's cotton spinning capacity, spoke enthusiastically about their own deals with local chief farmers and other "contractors", mostly former Licensed Buying Agents for the Cotton Board (Ismaila, interview 22/1 1987). Malam Abubakar of the NTMA, on the other hand, pointed an accusing finger at all those members of the Association, including the UNTL, who failed to contribute to the financing of NCC purchases (Abubakar, interview 29/1). He also accused members for lack of patriotism as they imported cotton. He blamed it on the fact that most companies, unlike his own, had foreign owners who could not be trusted in this respect. The recourse to imports is less surprising if it is true, as suggested by Abubakar, that the difference in prices at the factory gate was not much. While SFEM (see below) had raised the cost of imports drastically, world market prices were simultaneously at a rock bottom. At the time of the interview, imported cotton was about six naira per kg while the local one was at least five. In his view, this differential was too small to prevent unpatriotic firms from importing, especially as domestic supply was uncertain.

Members of the NTMA showed little confidence in the ability of the Association to ensure their raw material needs. In his memorandum, Abubakar (1987) appealed to the Federal Government to prohibit individual textile firms to engage in cotton buying and to enforce the position of NTMA's own cotton company as the sole buyer on behalf of the entire textile industry.

Government policies continued to vacillate, both at the federal and the state level. This came out most strikingly in the announcement by the Federal Government in early 1987 of a decision to go back on the policy of dissolving the marketing boards (Business Concord 16/1 and 20/1 1987). Reference was made by the Minister to the need to protect the employment of the tens of thousands of workers working with the boards. They had been laid off at the end of 1986 but were now called back. The Government suggested that the boards should be retained but privatised by capitalising assets and converting them into equity. The original position that they should be abolished outright was, however, reasserted after a few weeks' confusion (Business Concord 3/2 1987). There was speculation that the new about-turn was caused by "strong pressures from certain quarters", presumably including the World Bank that was keeping watch over Nigeria's implementation of the "Structural Adjustment Programme".

With the abolition of the boards, the political responsibility for marketing arrangements had shifted to the state level. The intention of the international sponsors of the move had no doubt been that there should be as little state intervention as possible. This, however, was not how it was seen by the state governments. Intervention was rampant and often contradictory. Most of the governments in the cotton growing states insisted on controlling the trade, including the demand that all buyers should be registered and licensed by the state. Kaduna State sought to restrict purchases to a small number of organisations, including its own Farmers' Supply Company. The Governor promised to ensure that no "middleman" would be allowed to buy cotton. An edict prohibited the sale of cotton outside "approved markets" with up to one year's imprisonment for culprits (including the forfeiting of vehicles to the state). The Kaduna Government also decreed that cotton must be sold at the government-guaranteed price (New Nigerian 11/10 and 31/10 1986). There was an attempt to prevent cotton from being traded across state boundaries. The Kaduna Commissioner of Education, on behalf of the Governor, suggested that it would be "unpatriotic" to sell outside the state (New Nigerian 6/11 1986).

All of this was far from the liberal intentions behind the abolition of the boards. In the end state governments became heavily involved in both direct purchasing and in financing crop purchases by other organisations. In Kaduna, the government spent 14 million naira on such purchases only to find that the manufacturers were reluctant to buy because of either imports or own buying arrangements (New Nigerian 28/2 and National Concord 27/2 1987). The experience of the 1986/87 cotton season does not quite allow us to evaluate the implications for the textile industry of marketing without marketing boards. Things were highly unsettled. It remains to be seen what new forms of marketing with what level of state involvement will work themselves out over the next few years. It is likely that the state will continue to play a major role in financing crop purchases and that this will remain an area of state patronage of political significance. State involvement is likely to be required in seed development, seed treatment and seed distribution. However, it is also likely that the forces behind the abolition of the boards will seek to prevent the direct participation by the states in marketing which was still so much in evidence in the first season.

To sum up. From the perspective of the relations between industry and agriculture, the most dramatic feature of the 1986/87 season was the abortive attempt by the manufacturers' association to take over the monopolistic functions of the defunct board. A major reason for the failure of the bid was the weakness of the NTMA itself and the commitments of leading members to pursue their own strategies. In the case of the single most important group of companies, around the UNTL, as discussed above, it had decided to go it alone, partly aspiring to model itself on the success story of the Nigerian Tobacco Company and the contractual relations with groups of individual farmers established by the latter. Other companies, like Afprint and the NTM had already committed themselves heavily to direct investment in cotton production. As suggested in the discussion of Afcott above the plantation strategy might well serve as a platform for contract farming and other commercial relations.

It also worked the other way: Participation by textile firms in cotton trading opened up avenues for new industry-based farmers as in the case of Walid Jibrin, one of the Nigerian directors of UNTL, highly involved in the cotton buying programme, while entering into cotton farming in his own right. He had started to grow cotton on some 50 ha of his family land, for sale to the buyers from his own company (interview Jibrin 13/1 1987). Whatever balance of these forms of involvement may work itself out, the abolition of the marketing board has clearly forced industry to step up its intervention on all fronts. While "free" imports under the new foreign exchange arrangements may temporarily allow for some recourse to imported cotton, continued foreign exchange constraints and related "nationalist" tariff polices are likely to sustain the push into local sourcing.

SFEM AND THE ABOLITION OF IMPORT LICENSING

In October 1986, another major economic policy reform altered the raw material situation for the textile industry, as well as for the manufacturing sector as a whole. The import licensing system was abolished after protracted pressures from the IMF and the World Bank as part of a general structural adjustment programme (Bangura 1987). Foreign exchange was now to be auctioned on a weekly basis by the Central Bank to accredited buyers, mostly commercial banks. The new system was called SFEM, the Second-tier Foreign Exchange Market, indicating that foreign exchange was also made available, at least for the time being, at a more favourable fixed rate (the "First tier") for limited purposes, mostly government debt payments. The two tiers were merged in mid-1987.

The coming of SFEM opened up a new traumatic phase in Nigeria's experience of crisis management. The price of foreign exchange rose by some 300-400 per cent. The full implications for Nigerian industry remain to be explored.[9] The impact on the agro-industrial link and backward integration was ambiguous.

Industrialists were free to import whatever they wished as long as they could raise enough naira to buy foreign exchange. Clearly, imports under the new deal became very expensive which offered fresh incentives for domestic sourcing of raw materials. Yet, the struggle for import licence in the past had had its own heavy cost, including the cost of uncertainty and delay. The effective access to imports at pre-SFEM exchange rates had therefore been very restricted. While local raw materials became less expensive relative to imported ones as a result of the devaluation, the price advantage of local produce was by no means given, especially as world market prices were exceptionally depressed for a wide range of commodities, including cotton. There was also the usual advantage of imports with respect to reliability and regularity of supplies. The inflationary impact of SFEM on domestic production will also have to be taken into account. Nigeria's new industrial farmers had to pay more for their imported machinery and chemicals.

The incentive or disincentive to domestic sourcing caused by SFEM will vary from one industry to another, depending on prices and cost differentials as well as on market structure. Some manufacturers will be able to pass on higher import costs to customers and therefore feel unconstrained in drawing on SFEM. Others will not. The general logic behind the movement into domestic sourcing and backward integration, however, is still at work: The sharp decline in supply of foreign exchange to the economy as a whole. SFEM was another way of rationing this scarce resource. For Nigerian industry, compelled to accommodate to a situation of scarcity, domestic sourcing remained an avenue for activating under-utilised capacity and maximising returns to the capital that had been locked up in idle plants.

The underlying assumption behind SFEM and the liberalisation of imports was that domestic production should be made competitive at levels of prices and costs as dictated by the post-devaluation value of the naira. As we saw above, the leadership of the NTMA was worried that, with world market prices at rock bottom levels, domestic cotton would not be competitive, unless manufacturers were patriotic enough to buy Nigerian cotton anyway (Abubakar, interview 29/1 1987).

A focus of struggle in the next few years is likely to be the level of tariff protection to be granted to domestic cotton. On the side of low tariffs we will find the "unpatriotic" industrialists as well as the liberal reformers who want to make sure that Nigerian producers become more cost effective through exposure from external competition. The high tariff lobby, however, may well prevail. All levels of the state as well as the organs of the textile industry have been engaged in an intense propaganda campaign for more cotton to be grown. It will be politically embarrassing if domestic producers at this point are allowed to be undercut by imports. It is therefore likely that the state will deviate from the liberal intentions behind the foreign exchange reform and grant protection to domestic cotton, at least in line with its own recommended prices. This is the more so, as the state has already attracted the anger of the cotton farmers because of the losses caused to them by the disruptive manner in which the Cotton Board was dissolved without any effective new arrangement to take its place. The anger of the farmers is likely to have been heard high up, especially as the peasants on this occasion were joined by people with more powerful voices, including not just our new industrialists-gone-farmers but a wider stratum of capitalist farmers, military officers, senior civil servants and businessmen with access to the corridors of power.

Why not ban the importation of cotton altogether? It would be consistent with government policy on other major agricultural imports such as rice, maize, and wheat. Although the dust raised by the chaotic 1986/87 cotton season remains to settle, the sharp increase in output suggests that national self-sufficiency is a possibility, even if the question-marks are many, especially on the marketing side. One problem is how to prevent manufacturers from escaping into synthetics. The level of protection that is established must reflect an "appropriate" balance between cotton and man-made fibres. We return to that option in the concluding chapter.

VI. Conclusions

The relationship between industry and agriculture in Nigeria is moving into a new phase. Industrial capital enters agriculture both directly through plantations and in various schemes for closer control over peasant and small capitalist farming. A deepening foreign exchange crisis has made investment in local sourcing a necessity in order to activate idle capacity at the industrial end. We see evidence of a process which is not specific to the textile industry nor to Nigeria. There is, we argue, a logic in import-dependent industrialisation that moves industry into a raw material trap, a dead end where it is compelled to venture into local sourcing in order to liberate profit potentials that have been locked up in idle machines. In the Nigerian case, the collapse of the oil economy only served to speed up such a process. We believe that the pressures towards backward integration would have developed anyway.

Even without instability in export earnings, import-dependent industry finds it difficult to sustain itself. It tends to generate more demands for imports than it replaces. Management fees, transfer of profits, etc. make additional demand on foreign exchange. The impact on the balance of payments is negative. This has been frequently observed in the literature on development. To radical critics it is seen as an important aspect of the underdevelopment syndrome and a case for a more self-centered development model.

What our study suggest and what seems less commonly realised is that such import-dependent industry may be compelled to attempt a break-out form this impasse. The direction taken by such an attempt will be determined by the balance of social and political forces specific to a particular society. Our study seeks to identify the particular forces that in the Nigerian case push the textile industry into the establishment of cotton plantations and discusses the developmental implications of such a move. For its realisation it attracts both the prompting by and the support of the state. While the basic logic is economic, the assistance of the state is required. Of critical importance is therefore an understanding of the social and political forces, domestic and international, that impress themselves on the administration of state power.

The main thrust of current initiatives - large-scale direct investments - is likely to reproduce import dependence in a more complex pattern of agro-industrial integration. Underdevelopment will be reproduced at a higher level (Beckman 1983). Foreign exchange continues to be a basic constraint that limits the scope of advance on such lines. It invites discrimination in the allocation of scarce resources in favour of the plantation owners at the expense of the peasant economy. The power of the state is required to remove the obstacles which stand in the way of capital in its effort to subordinate peasant land and labour.

As our case study of Afcott illustrates, it may not be expectations of high levels of productivity that determine decisions to invest in plantations. In a situation of extreme shortage, firms are anxious to establish direct control over supplies in order to escape from cut-throat competition in a sellers' market. Plantations can therefore pay their way not because they are more productive but because they allow parent companies to hold down procurement costs. Simultaneously, the new industrial farmers enjoy preferential access - backed and subsidised by the state -

to land and advanced production inputs. This gives plantations an advantage over small producers which does not presuppose any genuine productivity gains.

The move into plantations reflects the restructuring of production and social forces both nationally and internationally. Let us summarise some of the features.

AGRI-BUSINESS: A NEW TRANSNATIONAL PARTNERSHIP

The push towards direct investment by industry in agriculture links up with a new stage in the development of agro-industrialisation in the world economy. Accumulation in advanced capitalist agriculture and in related machinery and input-producing industries has reached a stage where pressures for internationalisation are intense. Transnational agri-business and their consultancy scouts offer their services aggressively in new markets, Nigeria being one. They sell packages of management, technology and inputs. They are not keen to invest directly because they fear confrontation with pre-capitalist social and political forces and they fear being trapped, without being able to remit profits from direct investments. They are therefore anxious to link up with local state and private companies who can take the financial risks and serve as effective partners in dealing with the authorities and with peasants and workers.

The Nigerian textile firms provide ideal partners in this respect. Many are part of transnational conglomerates but well entrenched within the local political economy. To the local industrialists, on the other hand, cooperating with foreign agri-business has its own attraction, especially as long as the marriage has state support. It allows them to bypass troublesome peasants. Agri-business with their marvellous machines can be brought in (duty free and tax deductable) to do the job.

Two tendencies in the development of the contemporary world economy thus come together to ensure mutual interests: industrial capital, caught in an import-substitution trap, looking for avenues of escape, and international agri-business in search of reliable partners in potentially important but risky new markets.

The political potential of this new partnership, however, depends on the ability of the agro-industrialists to join hands with such forces within the domestic ruling class that are capable of delivering political support. This is were the rising domestic agrarian bourgeoisie, with its roots and connections in the state apparatus, is of crucial political importance. The agro-industrialists add in their turn a significant new element to that class. Much depends on the ability of such a constellation of class forces to provide the legal, institutional, financial, and ideological backing for the plantation drive. The Nigerian state has gone a long way in recent years to open up the rural economy for agrarian capitalism, both domestic and transnational. But as we have seen, much more is expected from it, as in the case of the clamour for a revision of the Land Use Act to facilitate land acquisition.

CONSTRAINTS ON THE PLANTATION STRATEGY

The constraints are first of all political. The plantation strategy suggests the accelerated polarisation of class forces. It meets with resistance which may not be easily overcome, even when backing is provided by the repressive powers of the state, as illustrated in the case of the 1979-80 peasant rebellion at Bakolori, a state-sponsored irrigation project (Beckman 1985b). Peasant resistance may not only wreck fine feasibility calculations but also add political costs of unpredictable dimensions.

Plantations, in most cases, presuppose the appropriation of land previously at the disposal of the peasantry. It also requires the transformation of peasants into wage labourers. Plantations have to compete for labour in a rural economy where wage labour relations are marginal and where most producers have independent access to some means of production. The scope for suppressing wage levels on the plantations is therefore limited. The workers do not have to succumb to the dictates of capital as they have escape routes. Unlike small independent producers, the plantation companies will find it difficult to dodge minimum wage legislation or to bar unions from agitating among the workers.

Constraints on plantation agriculture are well known, not the least from the Nigerian experience (Williams 1986). The seasonality of the labour process poses problems in building an effective labour force and in holding down labour costs. The efficient utilisation of machinery is difficult, especially in a situation where maintenance and the supply of spare parts are irregular. Plantation management is also a scarce resource.

Such constraints may be overcome in the long run. In a forseeable future, the dependence on imported machinery and, as in Afcott's case, imported management will provide an additional and decisive foreign exchange constraint which prevents the plantation model from being more generally applied as a solution to the raw material problems of Nigerian industry.

How much cotton is likely to come out of the plantations? This is too early to tell. The foreign exchange constraint in particular, however, is likely to restrict the hectarage even if the political and managerial obstacles are overcome. A significant impact on the raw material gap will only be achieved if plantations are given high priority in the allocation of scarce resources of foreign exchange and management. From the point of view of the national economy, an attempt to generalise the plantation strategy as a solution to the raw material crisis is bound to lead to an inefficient use of such scarce resources. It also presupposes reliance on forms of production and technology which cannot be generalised at this particular stage of development. It generates another set of monopoly arrangements where profits and accumulation depend on preferential access ensured by the state. Because of the foreign input intensive nature of plantations profits will tend to gravitate away from the local as well as the national economy. Resources are simultaneously withheld from uses which would help to mobilise the basic productive assets of the society, above all human labour. Plantations not just fail to harness the vast productive potentials of the peasant population. They actively discriminate against them.

ALTERNATIVE SYSTEMS OF CONTROL

Investment in plantations is the most conspicuous approach chosen by the Nigerian textile industry in its efforts to grapple with the raw material crisis. It is not the only one. There are also attempts to design systems for closer control over independent producers. The Nigerian Tobacco Company (NTC) has practiced this with some success over a long period. We saw how UNTL, a leading textile firm, has plans to move in the same direction. Such schemes may also involve various combinations of "nucleus farms" (plantations) and packages of services to independent producers or outgrowers. The schemes may vary in the relative size of the plantation component, in the composition of the packages offered to outgrowers and in the control exercised by the companies over the producers (cf. Andrae 1983).

Plantations and contract farming are not mutually exclusive. The combinations can be placed on a continuum of control, ranging from "free" contracts at one end involving little direct participation by the company to the complete subordination of producers at the other end, including the transformation of independent producers into tenants on company land. In the case of Afcott, we saw how the outgrower component was only marginal to the plantation. The balance may well be shifted in favour of the outgrowers when investment funds for the plantation dry up. Some of the other companies which have acquired land may be more interested in extracting a surplus from tenants than to enter into direct production themselves. We also noted, in the case of Afcott, that the plantation also serves as a platform for offering services to other large producers in the area, securing a claim on future crops in return.

The Lancashire textile manufacturers in the colonial period were denied the option to enter into direct cotton production in Nigeria. Instead they sought to solve their raw material needs by trading with the peasants, establishing ginneries and pushing their own cotton seed. In the late colonial period, direct produce trading was taken over by the state marketing boards. With the abolition of the boards in 1987, Nigeria's textile industry, which had no experience of cotton buying on its own, was suddenly placed into a position of responsibility for organising the trade in cooperation with state governments.

While some companies have opted for plantations, the industry as a whole remains as dependent as ever on an efficient system for purchasing cotton from a mass of small producers. As we saw from the wranglings of the 1986/87 buying season, the new system of buying has yet to find its form. It is apparent, however, that there are forces at work, both within the textile industry itself and in state governments, that are anxious to reestablish some regulated system, including price control and restrictions on the free movement and trading of the crop. Intervention by state and industry in marketing is likely to continue to have a major impact on crop development.

CAN THE PEASANTS DO IT?

We have argued in this study against plantations as a solution to the raw material problem of the Nigerian textile industry. Industrial plantations are costly in terms of resources which are scarce (foreign exchange, advanced technology and management) and they fail to mobilise resources that are plentiful and poorly utilised: peasant land, labour and ingenuity. If an alternative strategy based on small-holders is to deliver the goods the current problems and confusions on the marketing side must be overcome. For instance, an efficient system for the development and distribution of good quality seeds is a must. But is this enough? Can the peasants do it?

Peasant production has been exposed to heavy pressures, including loss of labour and rising production costs. The fact that peasant farmers abandoned cotton, however, cannot be taken as evidence of the decline of the peasant economy as such. We have argued elsewhere that the shortages and high prices which characterised the Nigerian economy in the late 1970s and early 1980s were neither evidence of decline nor of stagnation. The commercial demand for agricultural produce had been allowed to expand at a formidable rate as a result of the way oil money was spent and which had caused drastic increase in the non-agricultural population (Beckman 1987a, b). The deflation of the economy in the mid-1980s, in combination with favourable weather, has largely eliminated previous deficits. It has been possible to ban the importation rice, maize and wheat while simultaneously reporting falling food prices and large surpluses of some crops, maize in particular.

Will the peasant economy also be able to accommodate an energetic programme of local sourcing by the textile industry and other agro-based industries which have depended on imports? We believe so, which does not mean that we find such an accommodation unproblematic. This is not the place to offer designs for a peasant-based resurrection of the Nigerian cotton industry. Preliminary evidence from the chaotic 1986/87 buying season suggests that recovery may be less distant than previously imagined.

Whatever the difficulties ahead, the sustained expansion of the Nigerian economy will depend on the mobilisation of the productive capacities of millions of small holders. Solutions which divert resources from or discriminate against this mass of small producers are likely to be counter-productive. The growth of their income will provide the primary basis for expanding mass markets for manufactured goods. In seeking solutions to the raw material problem, the textile industry should also consider what strategy stands the best chance of contributing to the demand for its own products.

PETRO-CHEMICALS TO THE RESCUE?

Some hope has been placed on the development of a domestic petro-chemical industry capable of bailing out the textile industry from its present predicament. Let us conclude with a brief note on that prospect. Plans for an integrated petro-chemical industry have been in the pipe-line since the beginning of the oil boom. The date currently suggested for the start of production is 1992. The first phase would include the production of chips which can be used by the synthetic fibre producers. Nigeria already has some capacity for making synthetic yarn (Nichemtex, Lagos) which depends entirely on imports. Some fresh capacity in this direction is also being developed.

As we have seen, the balance in the use of cotton and man-made materials in Nigeria's textile industry was roughly 60 - 40 in the early 1980s. The proportion of man-made increased sharply during the first half of the 1970s and was given an additional boost with the decline in utilisation of domestic cotton. It is tempting for textile manufacturers to continue the shift into synthetics. Greater emphasis on an industrially produced raw material would presumably reduce the uncertainties and hazards connected with agriculture. The move into synthetics becomes another way of escaping from the social, political and economic constraints of the agro-link.

The petro-chemical industry is awarded a strategic role in the development thinking of Nigeria's national planners. The textile industry is clearly one of the potentially most important markets for this new high-technology industry. For the planners it must be tempting to seek to ensure the widest possible demand among the textile firms. State planners and development finance institutions may therefore encourage further investment in processing capacity at the textile end. It may be taken as a laudable case of integrated industrial development.

Some backward integration into the local production of raw material for man-made fibres is no doubt justified. There will continue to be a demand for synthetics from final consumers that can not be easily redirected into cotton fabrics. However, there is a strong case, we believe, to keep the proportion of petroleum-based fibres as low as possible. Petroleum is a non-renewable resource, cotton is a renewable one. Petroleum can earn foreign exchange that can be used for vital imports and payments of services which cannot be easily replaced domestically. The ambition of the textile industry (NTMA) to cut the proportion of man-made fibres should therefore be encouraged, and not on import-substitution grounds alone.

Moreover, while cotton holds the prospect of providing work and income for a vast number of rural producers, the production of chips for man-made fibres is a highly capital-intensive process, involving little labour. Even more than in the case of mechanised, large-scale plantations, it requires a sophisticated technology that will have to be imported. Much of the income generated will leave the country. The textile industry should, in this case as well, opt for solutions to its raw material problem that expand the market for its own products. A prosperous peasantry should be its number one target.

Notes

1. For general references on Nigerian development, see Williams 1976 and Kirk-Greene and Rimmer 1981. For recent developments, see *West Africa* magazine or *Economist* 1986. For overviews of industrial and agricultural policies, see Forrest 1981 and 1982. On agriculture, see also Watts 1987.
2. The Isiaku Rabiu Group of Companies, a Kano based commercial and industrial conglomerate, exhibited a pyramid of such tins of palm oil, imported from West Germany, at the Kaduna Industrial Trade Fair in 1980, leading thoughts, embarrassingly, to the famous groundnut pyramids of the past.
3. For references to the agricultural ventures of the breweries, see also *Business Times* 23/4, 30/4, 3/9 1984, and *Guardian* 24/4 1984, 16/7 and 21/7 1985.
4. Cotton *lint* is the *fibre* after the seeds have been removed by ginning from the raw *seed cotton*. Lint is usually measured in *bales* in the context of production and internal trade. We have converted bales to *tonnes* for convenient comparison with other fibres and with yarn. One average bale of lint is 181,4 kg.
5. See IAR, Cropping Scheme Meetings: Notes on the Cotton and Fibres Improvement Programme/Fibres Research Programme, various years. See also IAR, Report to the Board of Governors: Cotton and Fibres Programme, various years, and Faulkner 1974. For a useful recent discussion, see Clough 1986.
6. Clough (1986) suggests that problems of pest control in early planted cotton was another cause of late planting.
7. For references to competition from food crops, see, e.g., IAR 1982 (including the address by the Director of the Federal Department of Agriculture and the Vice-Chancellor of ABU). See also various press statements by the General Manager of Nigerian Cotton Board (e.g., *New Nigerian* 31/10 1981, *NCB* 1985b and *CAP* 1985a. For recent field data on relative returns, see Clough 1986.
8. Clough (1986) documents the changing financial arrangements of the Licensed Buying Agents.
9. The operation of SFEM was surveyed and some preliminary assessments offered in a National Conference on SFEM organised by the Economics Department, Ahmadu Bello University, Zaria 27-28 January 1987.

References

ABUBAKAR, M. (1987) Brief on the state of the cotton industry in the aftermath of the dissolution of the Nigerian Cotton Board. KTL, Kaduna 2/1.

AFCOTT (1986) Cotton project. Outgrowers' agreement. Ngurore.

AFPRINT (1981-85) Annual reports and accounts. Lagos.

AKE, C. (1981) *A political Economy of Africa.* Harlow: Longman.

AMALE, S. A. (1986) River basin development and the peasant and agrarian problem: The case of Upper Benue. M Sc Diss., Dept. of Pol. Science, ABU, Zaria.

AMIN, S. (1974) *Accumulation on a World Scale.* New York: Monthly Review Press.

---- (1976) *Unequal Development.* New York: Monthly Review Press.

ANDRAE, G. (1983) "Agro-based industry in Nigeria: Forms of subordination." Paper presented at Seminar on Peasants and Agriculture in Africa. Scandinavian Institute of African Studies, Uppsala.

---- and B. BECKMAN (1982) Agro-based industry, food, and underdevelopment in Nigeria. Project proposal and outline. Stockholm: Dept. of Human Geography, University of Stockholm.

--------(1984) "Labour and industrial crisis: The case of Nigerian textiles and cotton". *AKUT* 30. Uppsala.

--------(1985) *The Wheat Trap. Bread and underdevelopment in Nigeria.* London: Zed Books.

AUTA, D. (1982) "A look at made in Nigeria textile goods", 1 and 2. *New Nigerian* 11/5 and 19/5 1982.

BANGURA, Y. (1987) "IMF/World Bank Conditionality and Nigeria's Structural Adjustment Programme" In Havnevik, K.J. (ed), *The IMF and the World Bank in Africa.* Uppsala: Scandinavian Institute of African Studies.

BECKMAN, B. (1983) "Marxism and underdevelopment. A critique of Ake." Paper to the conference "Marx and Africa", Zaria. November.

---- (1985a) "Neo-colonialism, capitalism and the state in Nigeria". In Bernstein, B. and B. K. Campbell (eds.) *Contradictions of Accumulation in Africa.* Beverly Hills: Sage.

---- (1985b) "Bakolori: Peasants versus state and capital". *Nigerian Journal of Political Science* 1985, IV:1-2.

---- (1987a) "Public investment and agrarian transformation in northern Nigeria". In Watts, M. (ed.) *State, Oil and Agriculture in Nigeria.* Berkeley: Institute of International Studies.

---- (1987b) "Peasants and democratic struggles in Nigeria." Paper to AKUT conference on "Labour and democracy" Uppsala and Review of African Political Economy conference on "Struggles in Africa", Liverpool. September 1986. As revised. Zaria.

BEEDEN, P. (1976) "The feasibility of improved sole crop cotton production technology for the small-scale farmer in the Northern Guinea Savanna Zone of Nigeria". *Samaru Miscellaneous Papers,* 61.

BUCH-HANSEN, M. and H. SECHER MARCUSSEN (1982) "Contract farming and the peasantry: Case studies from Kenya". *Review of African Political Economy,* 23.

C A P (1985a) Cotton & Agricultural Processors Ltd., Cotton production in Nigeria. Memorandum submitted to the "Committee on Cotton Production in the 10 Northern States" 20/5 1985. Zaria.

---- (1985b), Same. Letter ZN. 356/85, 22/10 1985.

C B N (1985) Central Bank of Nigeria. *Monthly Report* May 1985.

CLOUGH, P. (1986) The Production and Marketing of Cash Crops and Grains in Northern Nigeria, 1985. A View Based on Village Research. Consultancy Report, Western Africa Regional Office, The World Bank, Washington.

COWLEY, E. J. (1966) "Development of the cotton growing industry in Nigeria with special reference to the work of the British Cotton Growing Association". *Empire Cotton Growing Review* 43:3.

DUFF, H. (1921) Cotton growing in Nigeria: A report by Sir Hector Duff to the committee on a tour undertaken in Nigeria, Feb.-July, 1921. London: Empire Cotton Growing Corporation.

ECONOMIST (1986) "After the ball. A survey of Nigeria". 3/5.
ETUK, E. G. a.o. (1985) "Improved input use by small-scale cotton farmers in Northern Nigeria". Research in progress. Institute for Agricultural Research, Ahmadu Bello University. Zaria.
FAULKNER, R. C. (1974) "Cotton in Nigeria: Research towards increased production" *Span* 17:1. Reprint in *Samaru Research Bulletin* 219.
F G N (1977) Federal Government of Nigeria. Commodity Boards Decree, No. 29. Lagos.
---- (1981) Same. *Fourth National Development Plan* 1981-1985. Lagos.
---- (1982) Same. Economic Stabilisation (Temporary Provisions) Act. Lagos.
FORREST, T. (1981) "Agricultural policies in Nigeria 1900-78". In Heyer, J., Roberts, P. and Williams, G. (eds.), *Rural Development in tropical Africa*. London: Macmillan.
---- (1982) "Recent developments in Nigerian industrialization". In Fransmann, M. (ed), *Industry and Accumulation in Africa*. London: Heinemann.
F O S (1982, 1983 and various years) Federal Office of Statistics. *Nigerian Trade Summary. Lagos.*
---- (1985) Same. *Economic and Social Statistics Bulletin*. Lagos.
HOPKINS, A. G. (1973) *An Economic History of West Africa*. London: Longman.
I A R, Institute for Agricultural Research, Ahmadu Bello University, Zaria. Cotton and Fibres Programme. Report to the Board of Governors on the Institute's Work. Various years. Zaria.
---- (1982) Same. First National Symposium on Cotton Production. Proceedings. Forthcoming. Zaria.
---- (1983, 1984) Same. Cropping Scheme Meeting. Fibre Research Programme. Zaria.
---- (1985) Same. Proposals for research projects on mechanised large-scale cotton production and long-staple cotton submitted to the Nigerian Textile Manufacturers Association (NTMA). Zaria.
IKOKWU M. A. (1985) "Synthetic fibre plant as an offshoot of a petrochemical complex". Paper presented at the symposium on Raw Materials for the Textile and Allied Industries, Federal Institute of Industrial Research 12/4 1985. Lagos: NNPC Petrochemicals.
ILO, O. (1983) "Paper presented by the Nigerian Textile Manufacturers' Association at the National Workshop on Raw Materials for Nigerian Industries". Lagos.
---- (1984) Interview with Mr O. A. Ilo, General Manager, Nigerian Textile Mills, *Nigerian Enterprise*, September 1984.
ITMF (1984) International Textile Manufacturers Federation, *International Cotton Industry Statistics*.
JONES, T. S. (1969) Cotton production in Nigeria and the possibility of a cotton estate to produce longer staple cotton. Report. Commonwealth Development Corporation. London.
KILBY, P. (1969) *Industrialisation in an Open Economy. Nigeria 1945 - 66*. Cambridge: Cambridge University Press.
KIRK-GREENE, A. and RIMMER, D. (1981) *Nigeria since 1970. A Political and Economic Outline*. London: Hodder and Stoughton.
KRIESEL, H. C. (1968) *Cotton Marketing in Nigeria*. East Lansing: Consortium for the Study of Nigerian Rural Development, Michigan State University.
LANGDON, S. (1981) *Multinational Corporations in the Political Economy of Kenya*. London: Macmillan.
LITTLE (1969) Arthur D Little, Inc., Technical Advisors, The Nigerian Cotton Textile Industry: Its Prospects and Problems. Lagos: Industrial Planning Division of the Federal Ministry of Industries.
MANFU (1985) Manfu International Ltd., Boosting Cotton Production - An Extension Approach. A feasibility study sponsored by United Nigerian Textiles Ltd. Kaduna.
NABUGUZI, G. E. (1986) Contract farming. A comparative study of Kenya, Tanzania and Nigeria. M. Sc. Diss., Dept. of Pol. Science, ABU, Zaria.
N A D C (1971) National Agricultural Development Committee. Report of the Study Group on Cotton and other Fibres. Lagos: Federal Department of Agriculture.
N C B (1978 - 1984) Nigerian Cotton Board. *Annual Reports*. Funtua.
---- (1979b) Same. "Development of cotton production in Nigeria". New Nigerian 21/2 1979 (advertisement).
---- (1982b) Same. Activities, Achievements and Hopes for the Future. Funtua.

---- (1984b) Same. Causes for the Decline of Cotton and some Measures to remedy the Cotton Situation. Funtua.
---- (1985a) Same. Agricultural Marketing and Pricing System in Nigeria with particular reference to Cotton. Memo 22/3 1985. Funtua.
---- (1985b) Same. Memorandum on Cotton Production. Funtua.
---- (1985c) Same. Memorandum to Nigerian Manufacturers' Association Sub-Committee on Local Growing of Raw Materials.
NIGERIAN ENTERPRISE (1984) "Companies on the move: Afprint, sign of quality textiles." *Nigerian Enterprise*, September. Lagos.
NISER (1983a) Nigerian Institute of Social and Economic Research. Report for the Federal Ministry of Industry on National Workshop on Raw Materials for Nigerian Industries, July 1983. Lagos.
---- (1983b) "Textiles, Leather and Rubber Products". A pre-print prepared for the National Workshop on Raw Materials for Nigerian Industries, July 1983. Lagos.
N L C (1985) Nigerian Labour Congress. *Towards National Recovery*. Lagos. (Also published in Business Concord 19/7, 26/7, 2/8, 9/8 1985).
NORMAN, D. W. (1974) and J. A. Hayward and H. R. Hallam, "An assessment of cotton growing recommendations as applied by Nigerian farmers". *Cotton Growing Review 51*. (Also as reprint in *Samaru Research Bullentin* 230).
N T M A (1985a) Nigerian Textile Manufacturers Association. Preliminary results of a survey of capacity at 100 per cent utilisation. Unpublished. September. Lagos.
---- (1985b) Local sourcing of raw materials. A substitution plan in answer to requests from Federal Ministry of Commerce and Industry. September. Lagos.
OCULI, O. (1984) "Multinationals in Nigerian Agriculture in the 1980s". *Review of African Political Economy*, 31.
ONITIRI, H. M. A. and D. Olatunbosun (eds.) (1972) *The Marketing Board System*. Ibadan.
POKRANT, R. J. (1982) The Survival of Indigenous Tailoring among the Hausa of Kano City, Nigeria. Ph D Diss., University of Cambridge.
TIFFEN, M. (1967) *The Story of Nigerian Cotton*. London.
---- (1976) *The Enterprising peasant: Economic development in Gombe Emirate*, North Eastern State, Nigeria, 1900 - 1968. London: Ministry of Overseas Development.
WATTS, M. (ed) (1987) *State, Oil and Agriculture in Nigeria*. Berkeley: Institute of International Studies.
WILLIAMS, G. (1976) "Nigeria: a political economy". In Williams, G. (ed), *Nigeria: Economy and Society*. London: Rex Collings.
---- (1985) "Marketing without and with marketing boards: The origins of state marketing boards in Nigeria". *Review of African Political Economy*, 34.
---- (1986) Why is there no agrarian capitalism in Nigeria? Oxford: St Peter's College.
WORLD BANK (1973) Appraisal Reports of Funtua, Gombe and Gusau Agricultural Development Projects. 3 volumes. Washington.
---- (1981) *Accelerated Development in Sub-Saharan Africa: An Agenda for Action*. Washington.
---- (1987) *World Development Report*. Washington.

Periodicals

BUSINESS CONCORD. Weekly. Lagos.
BUSINESS TIMES. Weekly. Lagos.
FINANCIAL PUNCH.Weekly. Lagos.
GUARDIAN. Daily. Lagos.
NATIONAL CONCORD. Daily. Lagos.
NEW NIGERIAN. Daily. Kaduna.
SUNDAY NEW NIGERIAN.Weekly. Kaduna.
TRIUMPH. Weekly. Kano.
WEST AFRICA. Weekly. London.

Interviews

ABUBAKAR, M., Managing Director, Kaduna Textiles Ltd. Kaduna 19/7 1985.

AJAYI, O. W., Controller R and D, Nigerian Industrial Development Bank, NIDB, Lagos 12/9 1985.

EBURAJOLO, V., Executive Secretary, Nigerian Textile, Garment and Tailoring Employers' Association, Kaduna. 15/11 1985.

ETUK, E. G., Agricultural Economist, Institute for Agricultural Research, Ahmadu Bello University, Zaria. 11 1985.

HANCOCK, D., Farm Manager, Afcott, Ngurore. 2/7 1986.

IBIE, Administrative Director, President Clothing Company, Lagos. 20/11 1985.

ILO, O. A., Managing Director, Nigerian Textile Mills Ltd., Lagos and Chairman, Nigerian Textile Manufacturers' Association, Lagos. 20/11 1985.

ISMAILA, M. S., Assistant General Manager, United Nigerian Textiles Ltd., Kaduna. 22/1 1987.

JIBRIN, W., Assistant General Manager, United Nigerian Textiles Ltd., Kaduna. 13/1 1987.

KANKIA, M., Assistant General Manager (Operations), Nigerian Cotton Board, Funtua. 16/10 1985.

KNAGGS, G. H., Project Manager, Afcott, Ngurore. 30/6, 1/7 1986.

MAIRA, B., Chief Executive, Afprint Nigeria Ltd. Lagos. 10/9 1985.

PAHUJA, O. P., General Manager/Director, Enpee Industries Ltd., Lagos. 9/9 1985.

SANTURAKI, A. M., Project Supervisor, Mayo-Ine Project, Upper Benue River Basin Development Authority, Ngurore. 2/7 1986.

UDUKWU, J. C., Assistant General Manager (Sales), Nigerian Cotton Board, Funtua. 16/9 1985.

WADU, U., Principal Marketing Research Officer, Nigerian Cotton Board. Funtua. 9/10 1985.

YOUNG-ITIYE, D., Executive Secretary, Nigerian Textile Manufacturers' Association, Lagos. 13/9, 21/11 1985.

Research Reports

Below you will find a list of Research Reports published by the institute. Some of the reports are unfortunately out of print. Xero-copies of these reports can be obtained at a cost of SEK 0:50,- per page.

1. Meyer-Heiselberg, R., *Notes from Liberated African Department in the Archives at Fourah Bay College, Freetown, Sierra Leone.* 61 pp. Uppsala 1967. (OUT-OF-PRINT)
2. Not published.
3. Carlsson, Gunnar, *Benthonic Fauna in African Watercourses with Special Reference to Black Fly Populations,* 13 pp. Uppsala 1968. (OUT-OF-PRINT).
4. Eldblom, Lars, *Land Tenure — Social Organisation and Structure.* 18 pp. Uppsala 1969. (OUT-OF-PRINT)
5. Bjerén, Gunilla, *Makelle Elementary School Drop-out 1967.* 80 pp. Uppsala 1969. (OUT-OF-PRINT)
6. Møberg, Jens, Peter, *Report Concerning the Soil Profile Investigation and Collection of Soil Samples in the West Lake Region of Tanzania.* 44 pp. Uppsala 1970. (OUT-OF-PRINT)
7. Selinus, Ruth, *The Traditional Foods of the Central Ethiopian Highlands.* 34 pp. 1971. (OUT-OF-PRINT)
8. Hägg, Ingemund, *Some State-controlled Industrial Companies in Tanzania.* A case study. 18 pp. Uppsala 1971. SEK 10,-.
9. Bjerén, Gunilla, *Some Theoretical and Methodological Aspects of the Study of African Urbanization.* 38 pp. Uppsala 1971. (OUT-OF-PRINT)
10. Linné, Olga, *An Evaluation of Kenya Science Teacher's College.* 67 pp. Uppsala 1971. SEK 10,-.
11. Nellis, John R., *Who Pays Tax in Kenya?* 22 pp. Uppsala 1972. SEK 10,-.
12. Bondestam, Lars, *Population Growth Control in Kenya.* 59 pp. Uppsala 1972. SEK 10,-.
13. Hall, Budd L., *Wakati Wa Furaha. An Evaluation of a Radio Study Group Campaign.* 47 pp. Uppsala 1973. SEK 10,-.
14. Ståhl, Michael, *Contradictions in Agricultural Development.* A Study of Three Minimum Package Projects in Southern Ethiopia. 65 pp. Uppsala 1973. SEK 10,-.
15. Linné, Olga, *An Evaluation of Kenya Science Teachers College.* Phase II 1970-71. 91 pp. Uppsala 1973. SEK 15,-.
16. Lodhi, Abdulaziz Y., *The Institution of Slavery in Zanzibar and Pemba,* 40 pp. Uppsala 1973. ISBN 91-7106-066-9. (OUT-OF-PRINT)
17. Lundqvist, Jan, *The Economic Structure of Morogoro Town.* 70 pp. Uppsala 1973. ISBN 91-7106-068-5. (OUT-OF-PRINT)
18. Bondestam, Lars, *Some Notes on African Statistics.* Collection, reliability and interpretation. 59 pp. Uppsala 1973. ISBN 91-7106-069-4. (OUT-OF-PRINT)
19. Jensen, Peter Føge, *Soviet Research on Africa.* With special reference to international relations. 68 pp. Uppsala 1973. ISBN 91-7106-070-7. (OUT-OF-PRINT)
20. Sjöström, Rolf & Margareta, *YDLC — A Literacy Campaign in Ethiopia.* 72 pp. Uppsala 1973. ISBN 91-7106-071-5. (OUT-OF-PRINT)
21. Ndongko,Wilfred A., *Regional Economic Planning in Cameroon.* 21 pp. Uppsala 1974. SEK 15,-. ISBN 91-7106-073-1.
22. Pipping-van Hulten, Ida, *An Episode of Colonial History: The German Press in Tanzania 1901-1914.* 47 pp. Uppsala 1974. SEK 15,-. ISBN 91-7106-077-4.
23. Magnusson, Åke, *Swedish Investments in South Africa.* 57 pp. Uppsala 1974. SEK 15,-. ISBN 91-7106-078-2.
24. Nellis, John R., *The Ethnic Composition of Leading Kenyan Government Positions.* 26 pp. Uppsala 1974. SEK 15,-. ISBN 91-7106-079-0.
25. Francke, Anita, *Kibaha Farmers' Training Centre.* Impact Study 1965-1968. 106 pp. Uppsala 1974. SEK 15,-. ISBN 91-7106-081-2.
26. Aasland, Tertit, *On the move-to-the-Left in Uganda 1969-1971.* 71 pp. Uppsala 1974. SEK 15,-. ISBN 91-7106-083-9.
27. Kirk-Greene, A.H.M., *The Genesis of the Nigerian Civil War and the Theory of Fear.* 32 pp. Uppsala 1975. SEK 15,-. ISBN 91-7106-085-5.
28. Okereke, Okoro, *Agrarian Development Programmes of African Countries.* 20 pp. Uppsala 1975. SEK 15,-. ISBN 91-7106-086-3.
29. Kjekshus, Helge, *The Elected Elite. A Socio-Economic Profile of Candidates in Tanzania's Parliamentary Election, 1970.* 40 pp. Uppsala 1975. SEK 15,-. ISBN 91-7106-087-1.
30. Frantz, Charles, *Pastoral Societies, Stratification and National Integration in Africa.* 34 pp. Uppsala 1975. ISBN 91-7106-088-X. (OUT OF PRINT)
31. Esh, Tina & Rosenblum, Illith, *Tourism in Developing Countries - Trick or Treat?* A Report from the Gambia. 80 pp. Uppsala 1975. ISBN 91-7106-094-4. (OUT-OF-PRINT)
32. Clayton, Anthony, *The 1948 Zanzibar General Strike.* 66 pp. Uppsala 1976. SEK 15,-. ISBN 91-7106-094-4.
33. Pipping, Knut, *Land Holding in the Usangu Plain.* A survey of two villages in the Southern Highlands of Tanzania. 122 pp. Uppsala 1976. SEK 15,-. ISBN 91-7106-097-9.
34. Lundström, Karl Johan, *North-eastern Ethiopia: Society in Famine.* A study of three social institutions in a period of severe strain. 80 pp. Uppsala 1976. ISBN 91-7106-098-7. (OUT-OF-PRINT)
35. Magnusson, Åke, *The Voice of South Africa.* 55 pp. Uppsala 1976. ISBN 91-7106-106-1. (OUT OF PRINT)
36. Ghai, Yash P., *Reflection on Law and Economic Integration in East Africa.* 41 pp. Uppsala 1976. ISBN 91-7106-105-3. (OUT-OF-PRINT)
37. Carlsson, Jerker, *Transnational Companies in Liberia.* The Role of Transnational Companies in the Economic Development of Liberia. 51 pp. Uppsala 1977. SEK 15,-. ISBN 91-7106-107-X.
38. Green, Reginald H., *Toward Socialism and Self Reliance. Tanzania's Striving for Sustained Transition Projected.* 57 pp. Uppsala 1977. ISBN 91-7106-108-8. (OUT-OF-PRINT)
39. Sjöström, Rolf & Margareta, *Literacy Schools in a Rural Society.* A Study of Yemissrach Dimts Literacy Campaign in Ethiopia. 130 pp. Uppsala 1977. ISBN 91-7106-109-6. (OUT-OF-PRINT)
40. Ståhl, Michael, *New Seeds in Old Soil.* A study of the land reform process in Western Wollega, Ethiopia 1975-76. 90 pp. Uppsala 1977. SEK 15,-. ISBN 91-7106-112-6.
41. Holmberg, Johan, *Grain Marketing and Land Reform in Ethiopia.* An analysis of the marketing and pricing of food grains in 1976 after the land reform. 34 pp. Uppsala 1977. ISBN 91-7106-113-4. (OUT-OF-PRINT)
42. Egerö, Bertil, *Mozambique and Angola: Reconstruction in the Social Sciences.* 78 pp. Uppsala 1977. SEK 15,-. ISBN 91-7106-118-5.

43. Hansen, Holger Bernt, *Ethnicity and Military Rule in Uganda.* 136 pp. Uppsala 1977. ISBN 91-7106-118-5. (OUT-OF-PRINT)
44. Bhagavan, M.R., *Zambia: Impact of Industrial Strategy on Regional Imbalance and Social Inequality.* 76 pp. Uppsala 1978. SEK 15,-. ISBN 91-7106-119-3.
45. Aaby, Peter, *The State of Guinea-Bissau. African Socialism or Socialism in Africa?* 35 pp. Uppsala 1978. ISBN 91-7106-133-9. (OUT-OF-PRINT)
46. Abdel-Rahim, Muddathir, *Changing Patterns of Civilian-Military Relations in the Sudan.* 32 pp. Uppsala 1978. SEK 15,-. ISBN 91-7106-137-1.
47. Jönsson, Lars, *La Révolution Agraire en Algérie. Historique, contenu et problèmes.* 84 pp. Uppsala 1978. SEK 15,-. ISBN 91-7106-145-2.
48. Bhagavan, M.R., *A Critique of "Appropriate" Technology for Underdeveloped Countries.* 56 pp. Uppsala 1979. SEK 15,-. ISBN 91-7106-150-9.
49. Bhagavan, M.R., *Inter-Relations between Technological Choices and Industrial Strategies in Third World Countries.* 79 pp. Uppsala 1979. SEK 15,-. ISBN 91-7106-151-7.
50. Torp, Jens Erik, *Industrial Planning and Development in Mozambique. Some Preliminary Considerations.* 59 pp. Uppsala 1979. ISBN 91-7106-153-3. (OUT-OF-PRINT)
51. Brandström, Per, Hultin, Jan & Lindström, Jan, *Aspects of Agro-pastoralism in East Africa.* 60 pp. Uppsala 1979. SEK 15,-. ISBN 91-7106-155-X.
52. Egerö, Bertil, *Colonization and Migration. A summary of border-crossing movements in Tanzania before 1967.* 45 pp. Uppsala 1979. SEK 15,-. ISBN 91-7106-159-2.
53. Simson, Howard, *Zimbabwe - A Country Study.* 138 pp. Uppsala 1979. ISBN 91-7106-160-6. (OUT-OF-PRINT)
54. Beshir, Mohamed Omer, *Diversity Regionalism and National Unity.* 50 pp. Uppsala 1979. ISBN 91-7106-166-5. (OUT-OF-PRINT)
55. Eriksen, Tore Linné, *Modern African History: Some Historiographical Observations.* 27 pp. Uppsala 1979. SEK 15,-. ISBN 91-7106-167-3.
56. Melander, Göran, *Refugees in Somalia.* 48 pp. Uppsala 1980. SEK 15,-. ISBN 91-7106-169-X.
57. Bhagavan, M.R., *Angola: Prospects for Socialist Industrialisation.* 48 pp. Uppsala 1980. SEK. 15:-.ISBN 91-7106-175-4.
58. Green, Reginald H., *From Südwestafrika to Namibia. The political economy of transition.* 45 pp. Uppsala 1981. SEK 20,-. ISBN 91-7106-188-6.
59. Isaksen, Jan, *Macro-Economic Management and Bureaucracy: The Case of Botswana.* 53 pp. Uppsala 1981, SEK 20,-. ISBN 91-7106-192-4.
60. Odén, Bertil, *The Macroeconomic Position of Botswana.* 84 pp. Uppsala 1981. SEK 20,-. ISBN 91-7106-193-2.
61. Westerlund, David, *From Socialism to Islam? Notes on Islam as a Political Factor in Contemporary Africa.* 62 pp. Uppsala 1982. SEK 20,-. ISBN 91-7106-203-3.
62. Tostensen, Arne, *Dependence and Collective Self-Reliance in Southern Africa.* The case of the Southern African Development Coordination Conference (SADCC). 170 pp. Uppsala 1982. ISBN 91-7106-207-6. (OUT-OF-PRINT)
63. Rudebeck, Lars, *Problèmes de pouvoir populaire et de développement. Transition difficile en Guinée-Bissau.* 73 pp. Uppsala 1982. ISBN 91-7106-208-4. (OUT-OF-PRINT)
64. Nobel, Peter, *Refugee Law in the Sudan.* With The Refugee Conventions and The Regulation of Asylum Act of 1974. 56 pp. Uppsala 1982. SEK 20,-. ISBN 91-7106-209-2.
65. Sano, Hans-Otto, *The Political Economy of Food in Nigeria 1960-1982.* A Discussion on Peasants, State, and World Economy. 108 pp. Uppsala 1983. ISBN 91-7106-210-6. (OUT-OF-PRINT)
66. Kjærby, Finn, *Problems and Contradictions in the Development of Ox-Cultivation in Tanzania.* 164 pp. Uppsala 1983. SEK 40,-. ISBN 91-7106-211-4.
67. Kibreab, Gaim, *Reflections on the African Refugee Problem: A Critical Analysis of Some Basic Assumptions.* 154 pp. Uppsala 1983. ISBN 91-7106-212-2. (OUT-OF-PRINT) (New edition available from Africa World Press, P O Box 1892, Trenton, NJ 08608, USA)
68. Haarløv, Jens, *Labour Regulation and Black Workers' Struggles in South Africa.* 80 pp. Uppsala 1983. SEK 20,-. ISBN 91-7106-213-0.
69. Matshazi, Meshack Jongilanga & Tillfors, Christina, *A Survey of Workers' Education Activities in Zimbabwe, 1980-1981.* 85 pp. Uppsala 1983. SEK 20,-. ISBN 91-7106-217-3.
70. Hedlund, Hans & Lundahl, Mats, *Migration and Social Change in Rural Zambia.* 107 pp. Uppsala 1983. SEK 20,-. ISBN 91-7106-220-3.
71. Gasarasi, Charles P., *The Tripartite Approach to the Resettlement and Integration of Rural Refugees in Tanzania.* 76 pp. Uppsala 1984. SEK 20,-. ISBN 91-7106-222-X.
72. Kameir, El-Wathig & Kursany, Ibrahim, *Corruption as a "Fifth" Factor of Production in the Sudan.* 33 pp. Uppsala 1985. SEK 20,-. ISBN 91-7106-223-8
73. Davies, Robert, *South African Strategy Towards Mozambique in the Post-Nkomati Period.* A Critical Analysis of Effects and Implications. 71 pp. Uppsala 1985. SEK 20,-. ISBN 91-7106-238-6.
74. Bhagavan, M.R. *The Energy Sector in SADCC Countries.* Policies, Priorities and Options in the Context of the African Crisis. 41 pp. Uppsala 1985. SEK 20,-. ISBN 91-7106-240-8.
75. Bhagavan, M. R. *Angola's Political Economy 1975-1985.* 89 pp. Uppsala 1986. SEK 30,-. ISBN 91-7106-248-3.
76. Östberg, Wilhelm, *The Kondoa Transformation. Coming to grips with soil erosion in Central Tanzania.* 99 pp. Uppsala 1986. SEK 30,-. ISBN 91-7106-251-3.
77. Fadahunsi, Akin, *The Development Process and Technology.* A case for a resources based development strategy in Nigeria. 41 pp. Uppsala 1986. SEK 30,-. ISBN 91-7106-265-3.
78. Suliman, Hassan Sayed, *The Nationalist Movements in the Maghrib.* A comparative approach. 87 pp. Uppsala 1987. SEK 40,-. ISBN 91-7106-266-1.
79. Saasa, Oliver S., *Zambia's Policies towards Foreign Investment.* The Case of the Mining and Non-Mining Sectors. 65 pp. Uppsala 1987. SEK 30,-. ISBN 91-7106-271-8.
80. Andræ, Gunilla and Beckman, Björn, *Industry Goes Farming.* The Nigerian Raw Material Crisis and the Case of Textiles and Cotton. 68 pp. Uppsala 1987. SEK 40,-. ISBN 91-7106-273-4.